Communications
in Computer and Information Science   258

Tai-hoon Kim Hojjat Adeli
Alfredo Cuzzocrea Tughrul Arslan
Yanchun Zhang Jianhua Ma Kyo-il Chung
Siti Mariyam Xiaofeng Song (Eds.)

Database Theory and Application, Bio-Science and Bio-Technology

International Conferences, DTA and BSBT 2011,
Held as Part of the Future Generation
Information Technology Conference, FGIT 2011
in Conjunction with GDC 2011
Jeju Island, Korea, December 8-10, 2011
Proceedings

 Springer

Volume Editors

Tai-hoon Kim
Hannam University, Daejeon, Korea
E-mail: taihoonn@hannam.ac.kr

Hojjat Adeli
The Ohio State University, Columbus, OH, USA
E-mail: adeli.1@osu.edu

Alfredo Cuzzocrea
University of Calabria, Cosenza, Italy
E-mail: cuzzocrea@si.deis.unical.it

Tughrul Arslan
Edinburgh University, Edinburgh, UK
E-mail: t.arslan@ed.ac.uk

Yanchun Zhang
Victoria University, Melbourne, VIC, Australia
E-mail: yanchun.zhang@vu.edu.au

Jianhua Ma
Hosei University, Tokyo, 184-8584, Japan
E-mail: jianhua@hosei.ac.jp

Kyo-il Chung
ETRI, Daejeon, Korea
E-mail: kyoil@etri.re.kr

Siti Mariyam
University of Technology Malaysia, Johor, Malaysia
E-mail: mariyam@utm.my

Xiaofeng Song
Nanjing University of Aeronautics & Astronautics, Nanjing, P.R. China
E-mail: xfsong@nuaa.edu.cn

ISSN 1865-0929　　　　　　　　　e-ISSN 1865-0937
ISBN 978-3-642-27156-4　　　　　　e-ISBN 978-3-642-27157-1
DOI 10.1007/978-3-642-27157-1
Springer Heidelberg Dordrecht London New York

Library of Congress Control Number: Applied for

CR Subject Classification (1998): I.2, H.3, H.4, H.2.8, F.1, I.4

Typesetting: Camera-ready by author, data conversion by Scientific Publishing Services, Chennai, India
Printed on acid-free paper
Springer is part of Springer Science+Business Media (www.springer.com)

Foreword

Database theory and application and bio-science and bio-technology are areas that attract many professionals from academia and industry for research and development. The goal of the DTA and BSBT conferences is to bring together researchers from academia and industry as well as practitioners to share ideas, problems and solutions relating to the multifaceted aspects of database theory and application and bio-science and bio-technology.

We would like to express our gratitude to all of the authors of submitted papers and to all attendees for their contributions and participation.

We acknowledge the great effort of all the Chairs and the members of Advisory Boards and Program Committees of the above-listed event. Special thanks go to SERSC (Science and Engineering Research Support Society) for supporting this conference.

We are grateful in particular to the speakers who kindly accepted our invitation and, in this way, helped to meet the objectives of the conference.

December 2011   Chairs of DTA 2011 and BSBT 2011

Preface

We would like to welcome you to the proceedings of the 2011 Database Theory and Application (DTA 2011) and Bio-Science and Bio-Technology (BSBT 2011) conferences held during December 8–10, 2011, at Jeju Grand Hotel, Jeju Island, Korea.

DTA focused on various aspects of advances in database theory and application, while BSBT focused on various aspects of advances in bio-science and bio-technology. These conferences provided a chance for academic and industry professionals to discuss recent progress in the related areas. We expect that the conferences and their publications will be a trigger for further related research and technology improvements in this important subject.

We would like to acknowledge the great efforts of the DTA 2011 and BSBT 2011 Chairs, International Advisory Board, Committees, Special Session Co-chairs, as well as all the organizations and individuals who supported the idea of publishing this volume of proceedings, including the SERSC and Springer.

We are grateful to the following keynote, plenary and tutorial speakers who kindly accepted our invitation: Hsiao-Hwa Chen (National Cheng Kung University, Taiwan), Hamid R. Arabnia (University of Georgia, USA), Sabah Mohammed (Lakehead University, Canada), Ruay-Shiung Chang (National Dong Hwa University, Taiwan), Lei Li (Hosei University, Japan), Tadashi Dohi (Hiroshima University, Japan), Carlos Ramos (Polytechnic of Porto, Portugal), Marcin Szczuka (The University of Warsaw, Poland), Gerald Schaefer (Loughborough University, UK), Jinan Fiaidhi (Lakehead University, Canada) and Peter L. Stanchev (Kettering University, USA), Shusaku Tsumoto (Shimane University, Japan), Jemal H. Abawajy (Deakin University, Australia).

We would like to express our gratitude to all of the authors and reviewers of submitted papers and to all attendees for their contributions and participation, and for believing in the need to continue this undertaking in the future.

Last but not the least, we give special thanks to Ronnie D. Caytiles and Yvette E. Gelogo of the graduate school of Hannam University, who contributed to the editing process of this volume with great passion.

This work was supported by the Korean Federation of Science and Technology Societies Grant funded by the Korean Government.

December 2011

Tai-hoon Kim
Hojjat Adeli
Alfredo Cuzzocrea
Tughrul Arslan
Yanchun Zhang
Jianhua Ma
Kyo-il Chung
Siti Mariyam
Xiaofeng Song

Organization

Steering Co-chairs

Tai-hoon Kim	GVSA and University of Tasmania, Australia
Wai-chi Fang	Nasa JPL, USA

General Co-chairs

Alfredo Cuzzocrea	ICAR-CNR and University of Calabria, Italy
Tughrul Arslan	Engineering and Electronics, Edinburgh University, UK
Wai-chi Fang	National Chiao Tung University, Taiwan
Yanchun Zhang	Victoria University, Australia

Program Co-chairs

Jianhua Ma	Hosei University, Japan
Kyo-il Chung	ETRI, Korea
Siti Mariyam	Universiti Teknologi, Malaysia
Xiaofeng Song	Nanjing University of Aeronautics and Astronautics, China
Tai-hoon Kim	GVSA and University of Tasmania, Australia

International Advisory Board

Saman Halgamuge	University of Melbourne, Australia
Joseph Kolibal	University of Southern Mississippi, USA
Philip Maini	University of Oxford, UK
Byoung-Tak Zhang	Seoul National University, Korea
Aboul Ella Hassanien	Cairo University, Egypt

Publicity Co-chairs

Muhammad Khurram Khan	King Saud University, Saudi Arabia
Aboul Ella Hassanien	Cairo University, Egypt

Program Committee

Alfredo Cuzzocrea
Anne James
Aoying Zhou
Asai Asaithambi
A.Q.K. Rajpoot
Adrian Stoica
Ajay Kumar IIT
Antonio Berlanga de
    Jesús
Arun Ross
Bob McKay
Chan Chee Yong
Chunsheng Yang
Damiani Ernesto
Daoqiang Zhang
David Taniar
Carlos Juiz
Cesare Alippi
Dana Lodrova
Davide Anguita
Dong-Yup Lee
Djamel Abdelakder
    Zighed
Emiran Curtmola
Fan Min
Feipei Lai
Fionn Murtagh
Emilio Corchado
Farzin Deravi
Francisco de Paz
Francisco Herrera
Gang Li
George A. Gravvanis
Guoyin Wang
Haixun Wang
Hans-Joachim Klein
Hiroyuki Kawano
Hiroshi Sakai
Hui Yang
Hujun Yin
Jason T.L. Wang
Jesse Z. Fang
Jia Rong

Jian Lu
Jian Yin
Jixin Ma
Joel Quinqueton
Joshua Z. Huang
Jun Hong
Junbin Gao
Jake Chen
Janusz Kacprzyk
Jason T.L. Wang
Javier Ortega-Garcia
Jesús García Herrero
Jim Torresen
Jongwook Woo
Jose Alfredo Ferreira
    Costa
José Manuel Molina
Juan Manuel C.
    Rodríguez
Kai-Ping Hsu
Karen Renaud
Kay Chen Tan
Kenji Satou
Keun Ho Ryu
Krzysztof Stencel
Kayvan Najarian
Kenji Mizuguchi
Kevin Daimi
Lachlan McKinnon
Ladjel Bellatreche
Laura Rusu
Lee Mong Li
Li Ma
Martin Drahansky
Mathew Palakal
Matthias Dehmer
Meena K. Sakharkar
Michal Dolezel
Michael E. Schuckers
Miguel Angel Patricio
Morihiro Hayashida
Li Xiaoli
Liangjiang Wang

Lusheng Wang
Longbing Cao
Lucian N. Vintan
Mark Roantree
Masayoshi Aritsugi
Miyuki Nakano
Nor Erne Nazira Bazin
Omar Boussaid
Ozgur Ulusoy
Pabitra Mitra Mitra
Pang-Ning Tan
Paolo Ceravolo
Peter Baumann
Piotr Wisniewski
Pong C. Yuen
QingZhong Liu
Radim Dvorak
R. Ponalagusamy
Rattikorn Hewett
Richi Nayak
Roselina Sallehuddin
Sabine Loudcher
Sanghyun Park
Sang-Wook Kim
Sanjay Jain
Saman Halgamuge
Sanaul Hoque
Sara Rodríguez
Sridhar Radhakrishnan
Stan Z. Li
Stephen Cameron
Suash Deb
Shu-Ching Chen
Shyam Kumar Gupta
Stephane Bressan
Tadashi Nomoto
Takeru Yokoi
Tan Kian Lee
Tao Li
Tetsuya Yoshida
Theo Härder
Tomoyuki Uchida
Toshiro Minami

Tutut Herawan
Tatsuya Akutsu
Tommaso Mazza
Tony Xiaohua Hu
Vasco Amaral
Veselka Boeva
Vicenc Torra
Vikram Goyal
Weining Qian
Wenjie Zhang
William Zhu

Waleed Abdullah
Wei Zhong
Witold Pedrycz
Xiaohua Hu
Xiao-Lin Li
Xuemin Lin
Yan Wang
Yang Yu
Yang-Sae Moon
Yaoqi Zhou
Yong Shi

Yu Zheng
Ying Zhang
Yiyu Yao
Yongli Ren
Yoshitaka Sakurai
Young-Koo Lee
Zhuoming Xu
Zeyar Aung
Zhenan Sun
Zizhong Chen

Table of Contents

A Heuristic-Based Decision Tree Induction Method for Noisy Data

Nittaya Kerdprasop and Kittisak Kerdprasop

Data Engineering Research Unit, School of Computer Engineering,
Suranaree University of Technology, 111 University Avenue,
Nakhon Ratchasima 30000, Thailand
{nittaya,kerdpras}@sut.ac.th

Abstract. Decision tree is one of the most popular tools in data mining and machine learning to extract useful information from stored data. However, data repositories may contain noise, which is a random error in data. Noise in a data set can happen in different forms such as wrong labeled instances, erroneous attribute values, contradictory or duplicate instances having different labels. The serious effect of noise is that it can confuse the learning algorithm to produce a long and complex model. Such distorted result is due to the attempt to fit every training data instance, including noisy ones, into the model descriptions. This is a major cause of overfitting problem. Most decision tree induction algorithms apply either pre-pruning or post-pruning techniques during the tree induction phase to avoid growing a decision tree too deep down to cover the noisy data. We, on the contrary, design a loosely coupled approach to deal with noise. Our noise handling feature is in a separate phase from the tree induction. Both corrupted and uncorrupted data are clustered and heuristically selected prior to the application of the tree induction module. We observe from our experiments that upon learning from highly corrupted data, our approach shows a better performance than the conventional decision tree induction method.

Keywords: Robust tree induction, Noisy data, Heuristics, Logic Programming.

1 Introduction

Decision tree induction is a popular method for mining knowledge from data by means of decision tree building and then representing the end result as a classifier tree. Popularity of this method is due to the fact that mining result in a form of decision tree is interpretability, which is more concern among casual users than a sophisticated method but lacking of understandability [6]. A decision tree is a hierarchical structure with each internal node containing a decision attribute, each node branch corresponding to a distinct attribute value of the decision node, and the class of decision appears at the leaf node [3]. The goal of building a decision tree is to partition data with mixing classes down the tree until each leaf node contains data instances with pure class.

T.-h. Kim et al. (Eds.): DTA/BSBT 2011, CCIS 258, pp. 1–10, 2011.
© Springer-Verlag Berlin Heidelberg 2011

When a decision tree is built, many branches may be overly expanded due to noise in the training data set. Noisy data contain incorrect attribute values caused by many possible reasons, for instance, faulty data collected from instruments, human errors at data entry, errors in data transmission [1]. If noise occurs in the training data, it can lower the performance of the learning algorithm [14]. The serious effect of noise is that it can confuse the learning algorithm to produce too specific model because the algorithm tries to classify all records in the training set including noisy ones. This situation leads to the *overfitting* problem [4], [8], [13]. General solution to this problem is a tree pruning method to remove the least reliable branches, resulting in a simplified tree that can perform faster classification and more accurate prediction about the class of unknown data class labels [4], [8], [10].

Most decision tree learning algorithms are design with the awareness of noisy data. The ID3 algorithm [9] uses the pre-pruning technique to avoid growing a decision tree too deep down to cover the noisy training data. Some algorithms adopt the technique of post-pruning to reduce the complexity of the learning results. Post-pruning techniques include the cost-complexity pruning, reduced error pruning, and pessimistic pruning [7], [11]. Other tree pruning methods also exist in the literature such as the method based on minimum descriptive length principle [12], and dynamic programming based mechanism [2].

A tree pruning operation, either pre-pruning or post-pruning, involves modifying a tree structure during the model building phase. Our proposed method is different from most existing mechanism in that we deal with noisy data prior to the tree induction phase. Its loosely coupled framework is intended to save memory space during the tree building phase and to ease the future extension on dealing with streaming data. We present the research framework and the detail of our methodology in Section 2. The prototype of our implementation based on the logic programming paradigm is illustrated in Section 3. Efficiency of our implementation on noisy data is demonstrated in Section 4. Conclusion and discussion appear as a last section of this paper.

2 Robust-Tree Induction: Framework and Methodology

Our proposed system has been named *robust-tree induction* to enunciate our intention to design a decision tree induction method with a noise tolerant property. The framework as shown in Figure1 is composed of the robust-tree component, which is the noise tolerant decision tree induction part, and the testing component responsible for evaluating the accuracy of the decision tree model as well as reporting some statistics such as tree size and running time.

Noise tolerance of our robust-tree induction method can be achieved through the selection of the representative data, instead of learning from each and every training data. These selected data are used further in the tree building phase. Training data are first clustered by clustering module to find the mean point of each data group. The data selection module then uses these mean points as a criterion to select the training data representatives. It is a set of data representatives that to be used as input of the tree induction phase. Heuristics have to be applied as a threshold in the selection step and as a stopping criterion in the tree building phase. The algorithms of a main module as well as the clustering, data selection, and tree induction modules are presented in Figures 2-5, respectively.

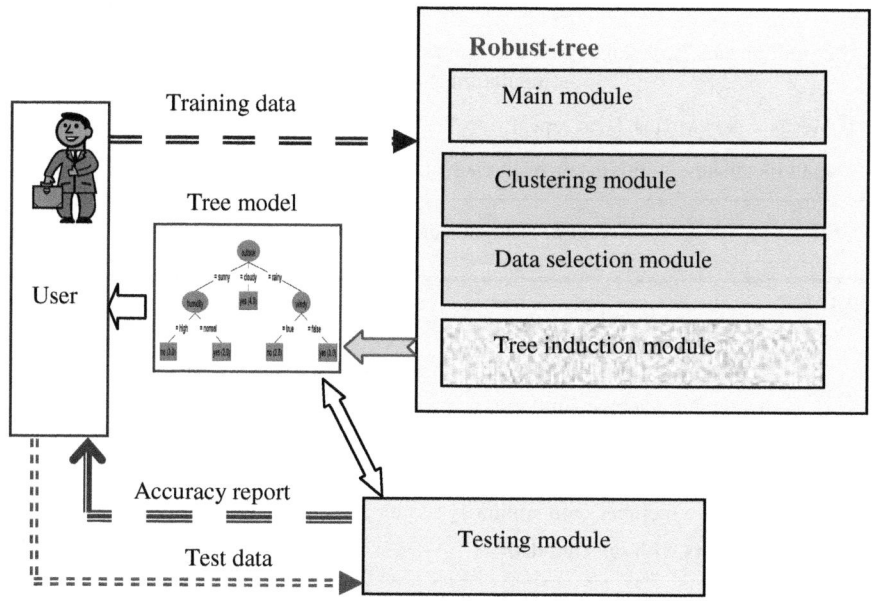

Fig. 1. A framework of the robust-tree induction system

Input: Data D with class label
Output: A tree model M
Steps:

    1. Read D and extract class label to check distinctive values K

    2. Cluster D to group data into K groups

    3. In each group

        3.1 Get mean attribute values

        3.2. Compute similarity of each member compared to its mean

        3.3 Compute average similarity and variance

        3.4 Set threshold $T = 2*Variance$

        3.5 Select only data with similarity > T

    4. Set stopping criteria S for tree building as

$$S = K - log \ [\ (number \ of \ removed \ data + K) \ / \ |D| \ ]$$

    5. Send selected data and criteria S into tree-induction module

    6. Return a tree model

Fig. 2. Main module of robust-tree induction system

Steps:

1. Initialize K means /* Create temporary mean points for all K clusters. */

2. Call find_clusters(K, Instances, Means) /* assign each data to the closest cluster;
    reference point is the mean of cluster */

3. Call find_means(K, Instances, NewMeans) /* compute new mean of each cluster; this
    computation is based on current members of each cluster */

4. If Means ≠ NewMeans Then repeat step 2

5. Output mean values and instances in each clusters

Fig. 3. Data clustering algorithm

Steps:

1. For each data cluster
2.   Compute similarity of each member compared to cluster mean
3.   Computer average similarity score of a cluster
4.   Computer variance on similarity of a cluster
5.   *Threshold = 2* variance*
6.   Remove member with similarity score < Threshold
7. Return K clusters with selected data

Fig. 4. Data selection algorithm

Steps:

1. If data set is empty   /* Case 1 */
2.   Then Assert(node(leaf,[Class/0], ParentNode)
3.   Exit   /* insert a leaf node in a database, then exit */

4. If number of data instances < MinInstances   /* Case 2 */
5.   Then Compute distribution of each class
6.   Assert(node(leaf, ClassDistribution, ParentNode)

7. If all data instances have the same class label   /* Case 3 */
8.   Then Assert(node(leaf, ClassDistribution, ParentNode)

9. If data > MinInstances and data have mixing class labels   /* Case 4 */
10.   Then BuildSubtree

11. If data attributes conflict with the existing attribute values of a tree   /* Case 5 */
12.   Then stop growing and create a leaf node with mixing class labels

13. Return a decision tree

Fig. 5. Tree induction algorithm.

3 System Implementation

The implementation of a robust-tree induction method is based on the logic programming paradigm using SWI-Prolog (www.swi-prolog.org). Data set takes the format of Horn clauses as illustrated in Figure 6.

```
attribute( outlook, [sunny, overcast, rainy]).
attribute( temperature, [hot, mild, cool]).
attribute( humidity, [high, normal]).
attribute( windy, [true, false]).
attribute( class, [yes, no]).
%
% data detail
%
instance(1, class=no, [outlook=sunny, temperature=hot, humidity=high, windy=false]).
instance(2, class=no, [outlook=sunny, temperature=hot, humidity=high, windy=true]).
instance(3, class=yes, [outlook=overcast, temperature=hot, humidity=high, windy=false]).
instance(4, class=yes, [outlook=rainy, temperature=mild, humidity=high, windy=false]).
instance(5, class=yes, [outlook=rainy, temperature=cool, humidity=normal, windy=false]).
```

Fig. 6. Part of a weather data set [11] displayed with the Horn clause format

Robust-tree induction system provides two level of noise tolerance: 0 and 1. Level 0 corresponds to ordinary ID3 style [9] without additional noise handling mechanism. Level 1 is a robust-tree induction with a heuristic-based mechanism to deal with noisy data. Prolog coding of noise tolerance levels 0 and 1 are presented as Prolog predicates $rtree(0, Attr)$ and $rtree(1, Attr)$, respectively as follows:

```
% ---------------------------------------
% start traditional tree-induction with ID3 algorithm

rtree(0, Attr) :-  !,
                        % make a list Ins = [1,2,...,n] of
                        % all instance ID
        findall(N, instance(N, _, _), Ins),
            % create decision tree, start with the root node
            % set MinInstance in leaf nodes = 1
            % then show model as decision tree
            % once finish building phase
        induce_tree(root, Ins, Attr, 1),
        print_tree_model.

%----------------------------------------------
% start clustering before induce a robust tree

    rtree(1, Attr) :-  !,
        attribute(class, ClassList),
        length(ClassList, K),
                % K is for specifying number of clusters
```

```
findall(N, instance(N,_,_),Ins),
clustering(Ins, K, Clusters, Means),
            % grouping instances
select_DataSample(Clusters,K,Means,[],Sample),
            % then select Sample
removed_Data(Sample, Ins, Removed),
length(Removed, R),
length(Ins, I),
MinInstance is K-log((R+K)/I),
            % a heuristic to prune tree
induce_tree(root,Sample,Attr,MinInstance),
print_tree_model.
```

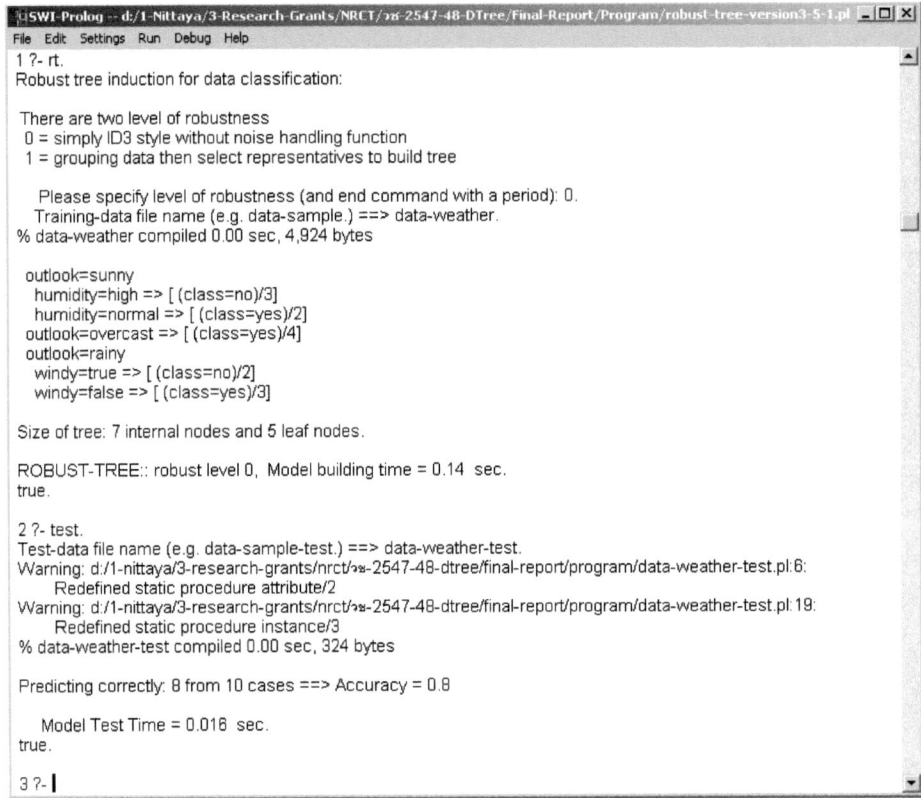

Fig. 7. User interface of the robust-tree induction system

4 Experimentation

On a series of experimentation, we compare size of the tree model as well as predicting accuracy of the robust-tree model with noise tolerance level 0 and 1. The tree model of weather data set with noise tolerance level 0 has been reported as follows:

outlook=sunny
    humidity=high
        windy=true => [(class=no)/1]
        windy=false
            temperature=hot => [(class=yes)/1]
            temperature=mild => [(class=no)/1]
            temperature=cool => [(class=yes)/0]
    humidity=normal => [(class=yes)/2]
  outlook=overcast => [(class=yes)/4]
  outlook=rainy
    windy=true => [(class=no)/2]
    windy=false => [(class=yes)/3]
Size of tree: 12 internal nodes and 8 leaf nodes.
ROBUST-TREE:: robust level 0, Model building time = 0.125 sec.

The tree model of the same data set, but the noise tolerance level has been set to be 1, shows a smaller tree as follows:

outlook=sunny
    humidity=high => [(class=no)/2, (class=yes)/1]
    humidity=normal => [(class=yes)/2]
  outlook=overcast => [(class=yes)/4]
  outlook=rainy
    windy=true => [(class=no)/2]
    windy=false => [(class=yes)/3]
Size of tree: 7 internal nodes and 5 leaf nodes.
Min instances in each branch = 3.94591
Initial Data = 14 instances
Removed Data = 0   instances
ROBUST-TREE:: robust level 1, Model building time = 0.0940001 sec.

For predicting accuracy comparison, user has to provide a separate test data set, which is also in a format of Horn clauses. Screenshot of predicting accuracy statistics of robust-tree can be displayed as in Figure 8.

ROBUST-TREE:: robust level 1, Model building time = 0.0940001 sec.
5 ?- test.
Test-data file name (e.g. data-sample-test.) ==> data-weather.
% data-weather compiled 0.00 sec, 0 bytes
Predicting correctly: 14 from 14 cases ==> Accuracy = 1
Model Test Time = 0.0 sec.

Fig. 8. Screen interface of the robust-tree induction system.

To test the accuracy and the noise tolerant efficiency of the proposed robust-tree induction system, we use the standard UCI data repository [5] including the monk, audiology, breast cancer, and vote data sets. Each data set is composed of a separate subset of training and test data. We prepare each training data set to contain eight levels of noise, that is, 0%, 1%, 5%, 10%, 15%, 20%, 25%, and 30%.

The comparison results of conventional decision tree induction algorithm (ID3 [9]) and the robust-tree induction algorithm in terms of model size and predicting accuracy are shown in Figures 9 and 10, respectively. It can be seen that robust-tree induction can produce a smaller tree model than conventional decision tree induction algorithm. The predicting accuracy of robust-tree model is higher than conventional decision tree model on most data sets, except the monk data set at the noise level 7-18% that the robust-tree model cannot outperform the conventional decision tree model.

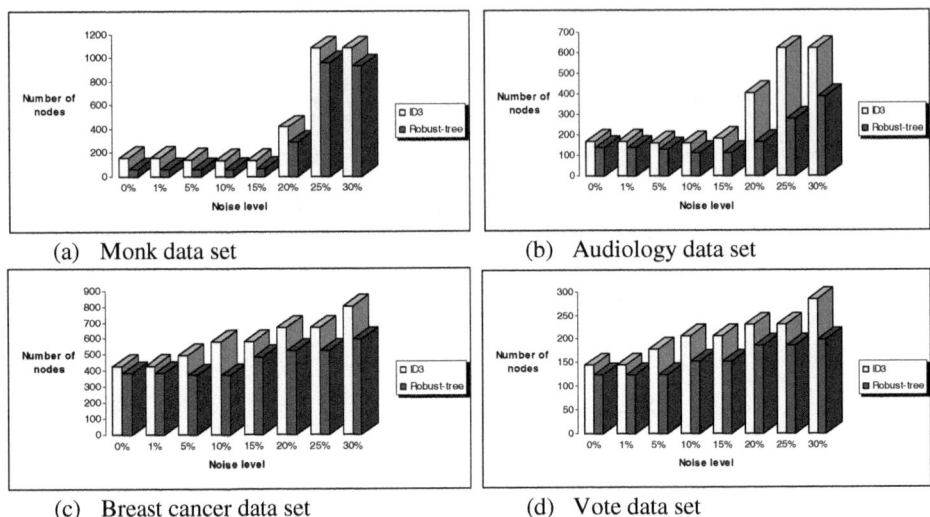

(a)   Monk data set   (b)   Audiology data set

(c)   Breast cancer data set   (d)   Vote data set

Fig. 9. Model size comparison of robust-tree induction against conventional induction

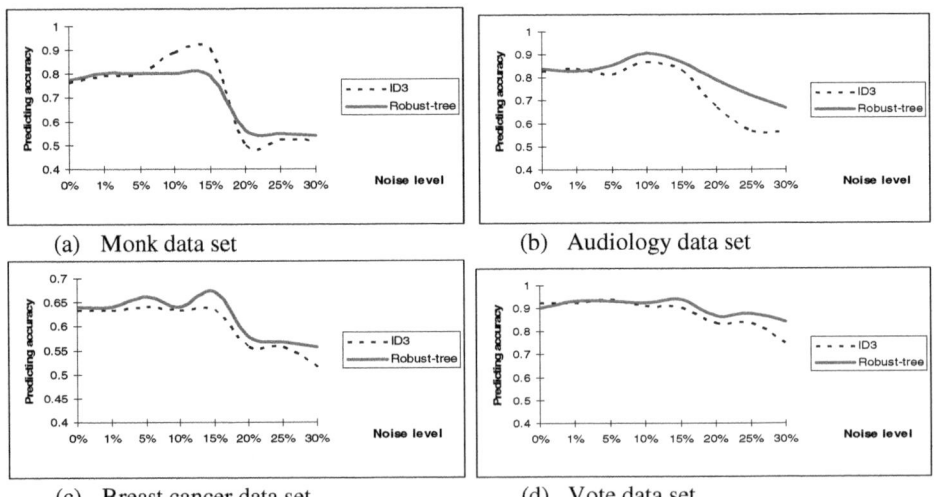

(a)   Monk data set   (b)   Audiology data set

(c)   Breast cancer data set   (d)   Vote data set

Fig. 10. Predicting accuracy of robust-tree induction against normal tree induction

On running time comparison, we record the time to build model in addition to the time for model testing (in seconds) of conventional decision tree induction against the robust-tree induction. Comparison results of the four data sets are reported in Table 1.

Table 1. Model building and testing time (in seconds) of the four data sets

	Noise level	Conventional decision tree induction	Robust-tree induction
Monk	0%	0.359+0.093=0.488	0.312+0.093=0.405
Data set	1%	0.344+0.141=0.485	0.344+0.094=0.438
	5%	0.343+0.110=0.453	0.297+0.063=0.36
	10%	0.360+0.094=0.454	0.375+0.094=0.469
	15%	0.313+0.110=0.423	0.375+0.078=0.453
	20%	0.766+0.250=1.016	0.797+0.25=1.047
	25%	2.609+0.954=3.563	2.593+0.968=3.561
	30%	2.594+0.859=3.453	2.656+0.906=3.562
Audiology	0%	0.414+0.069=0.483	0.398+0.041=0.439
data set	1%	0.409+0.092=0.501	0.387+0.044=0.431
	5%	0.383+0.077=0.460	0.379+0.038=0.417
	10%	0.397+0.095=0.492	0.361+0.046=0.407
	15%	0.471+0.133=0.604	0.375+0.033=0.408
	20%	1.577+0.181=1.758	0.884+0.059=0.943
	25%	1.965+0.911=2.876	1.726+0.656=2.382
	30%	2.018+0.965=2.983	1.998+0.701=2.699
Breast	0%	0.210+0.011=0.221	0.204+0.008=0.212
cancer	1%	0.197+0.015=0.409	0.213+0.015=0.228
data set	5%	0.274+0.048=0.322	0.186+0.011=0.197
	10%	0.211+0.069=0.280	0.197+0.020=0.217
	15%	0.298+0.053=0.351	0.214+0.029=0.243
	20%	0.323+0.079=0.402	0.231+0.037=0.268
	25%	0.465+0.93=1.395	0.240+0.056=0.296
	30%	0.512+0.104=0.616	0.443+0.077=0.520
Vote	0%	0.414+0.069=0.483	0.398+0.041=0.439
data set	1%	0.409+0.092=0.501	0.387+0.044=0.431
	5%	0.383+0.077=0.460	0.379+0.038=0.417
	10%	0.397+0.095=0.492	0.361+0.046=0.407
	15%	0.471+0.133=0.604	0.375+0.033=0.408
	20%	1.577+0.181=1.758	0.884+0.059=0.943
	25%	1.965+0.911=2.876	1.726+0.656=2.382
	30%	2.018+0.965=2.983	1.998+0.701=2.699

5 Conclusion

Noisy data can cause serious problem to many learning algorithms in terms of distorted results and the decrease in predicting performance of the learning results. In this paper, we propose a methodology to deal with noise in a decision tree induction algorithm. Our intuitive idea is to select only potential representatives, rather than applying the whole training data that some values are corrupted, to the tree induction algorithm.

Data selection process starts with clustering to form groups of similar data items in order to obtain the mean point of each data group. For each data group, the selection heuristic $T = 2 * Variance\_of\_cluster\_similarity$ will be used as a threshold to select only data around mean point within this T distance. Data that lie far away from the mean point are considered prone to noise and outliers; we thus remove them.

The removed data still play their role as one factor of a tree building stopping criterion, which can be formulated as $S = K - log[(number\ of\ removed\ data\ instances + K) / D]$, where K is the number of clusters, which has been set to be equal to the number of class labels, and D is the number of training data.

From experimental results, it turns out that our heuristic-based decision tree induction method is robust to data set with a high level of noise. It also produces a compact tree model. With such promising results, we thus plan to improve our methodology to be incremental such that it can learn model from steaming data.

Acknowledgments. This work has been supported by grants from the National Research Council of Thailand (NRCT) and Suranaree University of Technology via the funding of Data Engineering Research Unit.

References

1. Angluin, D., Laird, P.: Learning from noisy examples. Machine Learning 2, 343–370 (1988)
2. Bohanec, M., Bratko, I.: Trading accuracy for simplicity in decision trees. Machine Learning 15, 223–250 (1994)
3. Breiman, L., Freidman, J., Olshen, R., Stone, C.: Classification and Regression Trees. Wadsworth (1984)
4. Esposito, F., Malerba, D., Semeraro, G.: A comparative analysis of methods for pruning decision trees. IEEE Trans. Pattern Analysis and Machine Intelligence 19(5), 476–491 (1997)
5. Frank, A., Asuncion, A.: UCI Machine Learning Repository. School of Information and Computer Science. University of California, Irvine (2010), `http://archive.ics.uci.edu/ml`
6. Han, J., Kamber, H.: Data Mining: Concepts and Techniques, 2nd edn. Morgan Kaufmann (2006)
7. Kim, H., Koehler, G.J.: An investigation on the conditions of pruning an induced decision tree. European Journal of Operational Research 77(1), 82 (1994)
8. Mingers, J.: An empirical comparison of pruning methods for decision tree induction. Machine Learning 4(2), 227–243 (1989)
9. Quinlan, J.R.: Induction of decision tree. Machine Learning 1, 81–106 (1986)
10. Quinlan, J.R.: Simplifying decision tree. In: Gaines, B., Boose, J. (eds.) Knowledge Acquisition for Knowledge Based Systems, vol. 1. Academic Press (1989)
11. Quinlan, J.R.: C4.5: Programs for Machine Learning. Morgan Kaufmann (1992)
12. Quinlan, J.R., Rivest, R.: Inferring decision trees using the minimum description length principle. Information and Computation 80(3), 227–248 (1989)
13. Schaffer, C.: Overfitting avoidance bias. Machine Learning 10, 153–178 (1993)
14. Talmon, J.L., McNair, P.: The effect of noise and biases on the performance of machine learning algorithms. Int. J. Bio-Medical Computing 31(1), 45–57 (1992)

Optimizing Database Queries with Materialized Views and Data Mining Models

Nittaya Kerdprasop and Kittisak Kerdprasop

Data Engineering Research Unit, School of Computer Engineering,
Suranaree University of Technology, 111 University Avenue,
Nakhon Ratchasima 30000, Thailand
{nittaya,kerdpras}@sut.ac.th

Abstract. The process of intelligent query answering consists of analyzing the intent of a query, rewriting the query based on the intention and other kinds of knowledge, and providing answers in an intelligent way. Producing answers effectively depends largely on users' knowledge about the query language and the database schemas. Knowledge, either intentional or extensional, is the key ingredient of intelligence. In order to improve effectiveness and convenience of querying databases, we design a systematic way to analyze user's request and revise the query with data mining models and materialized views. Data mining models are constrained association rules discovered from the database contents. Materialized views are pre-computed data. This paper presents the knowledge acquisition method, its implementation with the Erlang programming language, and a systematic method of rewriting query with data mining models and materialized views. We perform efficiency tests of the proposed system on a platform of deductive database using the DES system. The experimental results demonstrate the effectiveness of our system in answering queries sharing the same pattern as the available knowledge.

Keywords: Query optimization, Materialized view, Data mining model, Query rewriting, Deductive database.

1 Introduction

Since the emergence of data mining as a new research area two decades ago, it has been argued among database researchers that the database management system (DBMS) should support both data processing and data mining tasks [3], [4], [5], [9]. With the data mining functionalities, a database can contain not only data contents and schemas, but also data models which are generalized information induced from the data contents. By providing such an extension framework of the DBMS, users can manipulate and access data models in the same manner as querying and processing the data contents. To achieve this aim, a number of database system extensions, such as the IBM intelligent miner [17] and Microsoft OLE DB for data mining [20], have been implemented.

Besides the front-end support for mining database contents, we propose that the next generation DBMS should also utilize the data mining models at the back-end part

T.-h. Kim et al. (Eds.): DTA/BSBT 2011, CCIS 258, pp. 11–20, 2011.

to support query answering and optimization. The purpose of query optimization is to rewrite a given query into an equivalent one that uses less time and resources. Equivalence is defined in terms of identical answer sets. Semantic knowledge such as integrity constraints and data mining models can be used to transform a query into an optimized one. Recent advance on semantic knowledge utilization has moved toward the setting of cooperative or intelligent query answering [11], [16].

The process of intelligent query answering consists of analyzing the intent of a query, rewriting the query based on the intention and other kinds of knowledge, and providing answers in an intelligent way [19]. Intelligent answers could be generalized, neighborhood or associated information relevant to the query. This concept is based on the assumption that some users might not have a clear idea of the database contents and schemas. Therefore, it is difficult to pose queries correctly to get some useful answers.

Knowledge, either intentionally or extensionally stated, is the key ingredient of intelligence. Many researchers [2], [7], [16], [19] propose to integrate data mining techniques as a knowledge discovery engine to serve an intelligent query answering purpose. We extend this idea by incorporating both data mining models and materialized views in the query answering system.

Data mining models are generalized rules discovered from databases and stored as tables, whereas materialized views are view relations computed and stored in the database as well. We consider data mining models and materialized views as semantic constraints capable of transforming queries to be processed intelligently.

We design a query optimizing and answering system using a platform of deductive database that can be easily integrated with induced knowledge obtained from the constrained association mining program implemented with the Erlang language. The remainder of the paper is organized as follows. Section 2 presents related work. Section 3 explains architecture and implementation of the proposed system. Section 4 discusses the experimentation and its results. Section 5 concludes the paper and indicates our future work.

2   Related Work

Evaluating queries efficiently and intelligently requires an important step of query rewriting and modification. Query rewriting is a basic step in query processing aiming at transforming a given query into another more efficient one that uses less time and resources to execute. A rewritten query normally produces the same answer set as the original query.

Query modification [7] interprets query rewriting in a more relaxing way as a query refining process to produce answers that might be a superset of the expected answers. The advantage of query relaxation is the increased possibility of obtaining desired answers when users have limited knowledge about the problem domain and the database schemas.

Early research in query modification [7], [11] has focused on rewriting the query using generalization concept, neighborhood, and type abstraction hierarchy. The work of Han et al [16] is among the early research in intelligent query answering that incorporates data mining techniques to rewrite users' queries. Their query relaxation approach employed the notion of generalization to build concept hierarchy.

Lin et al [19] proposed to integrate neighborhood information and data mining rules discovered from the databases to rewrite the queries. Muslea [21] introduced the LOQR algorithm to learn some knowledge about the problem domain using a small subset of the database. Then the learned information is used to relax the constraints in the query that originally returns an empty answer. Aragao and Fernandes [3] proposed a unified foundation for query answering and knowledge discovery. The combined system is called CIDS (Combined Inference Database Systems).

The integration of knowledge discovery and query answering system is also the basis of our research. However, we propose to extend the idea by incorporating not only the knowledge discovered from databases, but also the materialized views in the process of query rewriting and answering.

Materialized views are pre-computed data that are stored in the database. Answering queries using views has long been extensively studied [1], [6], [8], [10], [14], [15], [22]. Materialized views can provide useful information in query processing especially in the context of web searching applications. We thus design our system to employ both learned knowledge and materialized views to refine the given query.

3 Query Optimizer Design and Its Implementation

Our design of intelligent query answering system includes the semantic optimizer to utilize materialized views and data mining models as major sources of knowledge in semantically transforming the users' queries. The framework of the optimizer and its algorithm are provided in Figures 1 and 2, respectively.

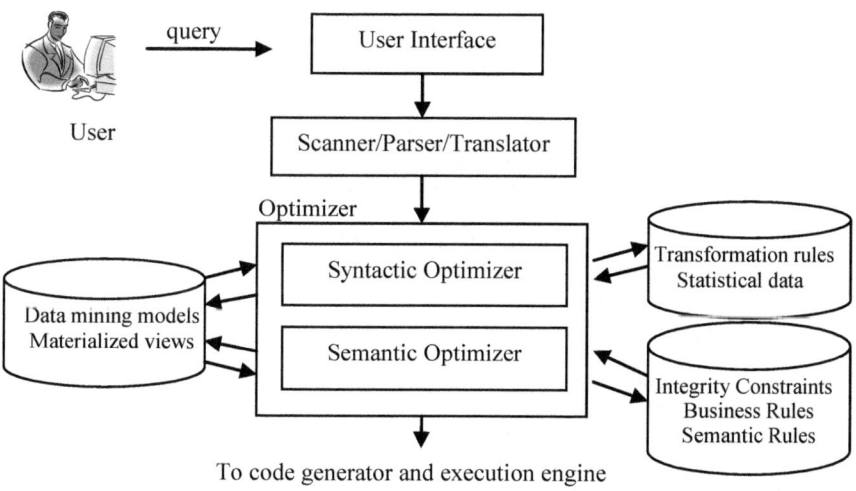

Fig. 1. A framework of query optimizer

Input: a database D,
                  a set of semantic rules S,
                  a set of materialized views V,
                  current user's query Q
Output: a new query Q′
Steps:

1.   Extract conditions C from the user's query Q
2.   For each $c \in$ C
3.   Search for applicable semantic rules from S by
3.1   assert c as a temporary database fact
3.2   search for predicates in S that are related to c
3.3   report searching result as an answer set A
4.   If A is empty, then return Q; otherwise proceed to the next step
5.   Form a new query Q′ by
5.1   Construct a head of query clause with C appeared as arguments
5.2   Construct clause body with applicable materialized view from V
5.3   Conjunct a clause body with predicates appeared in A
6.   Return a new query Q′

Fig. 2. Semantic query optimizer algorithm

Semantic rules as mentioned in the framework and in the algorithm are association rules [2] induced from the database contents and constrained to induce only knowledge relevant to users' queries. The association mining component is implemented with the Erlang programming language [13]. The program is written to search for association rules with confidence 1.0, that is, 100% correct. Some part of the source code can be displayed as follows:

```
% ------- a part to specify support value ----------------------------
inputSup(AllInput) ->
        Total = length(AllInput),
        {_,Per} = io:read(" input percent> "),   % read minimum support value
        MinSup = Total*Per/100 .
% ------- main function of association mining ---------------------
main() ->
      NameList = mylib:read_file( "ipum.NAMES"," ,.\t:|" ),
      mylib:text_file(write, NameList, filetemp),
      [H | Tail] = NameList,
      [F, C | T] = lists:reverse(H),
         %
         % ...... some codes are omitted here ........
      DB = myToSet(AllInput),
      MinSup = inputSup(AllInput),
      mylib:c(20, MinSup),
      mylib:text_file(write, AllInput, allinput),
      Items = my_flat(PossibleValue),
      AllL = apriori1(DB, Items, MinSup).
```

```
% --------- main1() is for creating rules ---------------
main1() ->
    {_, [AllL]}= file:consult("set.raw"),
    AllAsso2 = [list(X) || {X,_} <-AllL, length(list(X))>1 ],
    AllRuleGen = lists:flatten([genRule(L, length(L),
                            length(L)) || L <- AllAsso2]),
    AllRuleConf = [findConf(X, AllL) || X <- AllRuleGen],
    format("~nAllRule=~p ,~nThere are ~p rules ",
                            [AllRuleConf, length(AllRuleConf)]),
    mylib:text_file(write, AllRuleConf, allrule),
    Sorted = lists:sort(fun({_,_,C1},{_,_,C2}) ->  C1 >= C2 end, AllRuleConf),
    mylib:text_file(write, Sorted, "allsortedrule.txt"),
    Conf1= lists:filter(fun({A,B,C}) -> C==1.0 end, Sorted),  % confidence=1.0
    {_,IO} = file:open("rules.pl",[write]),  % write to file
    lists:map(fun(EachR) ->
                        transform_to_datalog(EachR, IO) end, Conf1),
    _ = file:close(IO).
```

4 Experimentation

On a series of experimentation, we draw a sample set containing 1,000 data records from the IPUMS-USA database which has been made available to public by Minnesota Population Center [18]. This database contains the United States socio-economic data of the year 1999 with household and person information as shown in Figure 3. The household records contain information such as value of household unit, monthly rental payment, family total income, number of families within each household unit, age of a person, relationship to household head, and other related information. Examples of household records are illustrated in Figure 4.

For the person records, the information is about education, employment status, occupation, income of a person, and other personal information. Examples of person records are shown in Figure 5.

Fig. 3. Variables as appeared in the household and person records [18]

table1(1,gq_1,gqtypeg_0,farm_1,ownershg_1,value_6,rent_0,ftotinc_3,nfams_1,
        ncouples_1,nmothers_0,nfathers_0,momloc_0,stepmom_0,momrule_0,
        poploc_0,steppop_0,poprule_0,sploc_2,sprule_1,famsize_2,nchild_0,
        nchlt5_0,famunit_1,eldch_10,yngch_10,nsibs_0,relateg_1,age_6,sex_1,
        raceg_1,marst_1,chborn_0,bplg_1).
table1(2,gq_1,gqtypeg_0,farm_1,ownershg_2,value_7,rent_9,ftotinc_3,nfams_2,
        ncouples_0,nmothers_1,nfathers_0,momloc_0,stepmom_0,momrule_0,
        poploc_0,steppop_0,poprule_0,sploc_0,sprule_0,famsize_2,nchild_1,
        nchlt5_0,famunit_1,eldch_1,yngch_1,nsibs_0,relateg_1,age_4,sex_2,
        raceg_1,marst_4,chborn_2,bplg_1).

Fig. 4. Household records containing 34 attributes

table2(1,school_1,educrec_7,schltype_1,empstatg_1,labforce_2,occscore_3,sei_2,
        classwkg_2,wkswork2_4,hrswork2_6,yrlastwk_0,workedyr_2,inctot_3,
        incwage_3,incbus_1,incfarm_1,incss_1,incwelfr_1,incother_1,poverty_6,
        migrat5g_1,migplac5_0,movedin_7,vetstat_1,tranwork_10,occupation_2).
table2(2,school_1,educrec_8,schltype_1,empstatg_1,labforce_2,occscore_2,sei_3,
        classwkg_1,wkswork2_3,hrswork2_6,yrlastwk_0,workedyr_2,inctot_2,
        incwage_2,incbus_2,incfarm_1,incss_1,incwelfr_1,incother_1,poverty_6,
        migrat5g_1,migplac5_1,movedin_0,vetstat_1,tranwork_10,occupation_3).

Fig. 5. Person records, each record has 27 attributes

To test the efficiency of query optimization with materialized views and data mining models, we create a database using the DES system [12]. The database stores table1 and table2 (Figures 4 and 5) to contain household and person information, respectively. Both tables are in the format of Datalog clauses. Data mining models used in the experimentation are association rules that are also transformed to be Datalog clauses as shown some part in Figure 6.

```
p(farm_1) :- p(gq_1), p(bplg_1).
p(farm_1) :- p(gqtypeg_0), p(bplg_1).
p(farm_1) :- p(migplac5_1), p(bplg_1).
p(gqtypeg_0) :- p(bplg_1), p(gq_1).
p(gq_1) :- p(gqtypeg_0), p(bplg_1).
p(farm_1) :- p(gqtypeg_0), p(eldch_10).
p(nchild_0) :- p(eldch_10), p(farm_1).
p(eldch_10) :- p(farm_1), p(nchild_0).
p(farm_1) :- p(nchild_0), p(eldch_10).
p(schltype_1) :- p(school_1), p(stepmom_0), p(farm_1), p(gq_1), p(gqtypeg_0).
```

Fig. 6. Part of semantic rules transformed from the association mining models

We test the query processing performance with six kinds of queries (experimental results are graphically shown in Figure 7 and summarized in Table I). Some queries (Query1, Query2, and Query3) cannot benefit from the presence of semantic rules because the queries' conditions do not fit the rules. For this case, the system takes more time to search for semantic rules than traditional direct querying method. But for some queries that fit the rule antecedents, the intelligent method does significantly save the database searching time. The details of each query as well as its running time report are provided as follows:

Query 1: Ask for value of farm housholds. (Null answer; all households are non-farm)
DES-Datalog> /assert query1(X) :- table1(A1,A2,A3,A4,A5,A6,A7,A8,A9,A10,A11,A12,
        A13,A14,A15,A16,A17,A18,A19,A20,A21,A22,A23,A24,A25,
        A26,A27,A28,A29,A30,A31,A32,A33,A34),A4=farm_2,X=A6.
DES-Datalog> query1(X).
        { }
        Info: 0 tuples computed. Total elapsed time: 110 ms.
Search for semantic rules to transform query: (No rules applied; waste time 295 ms.)
        DES-Datalog> /assert p(farm_2).
        DES-Datalog> p(C).
        { p(farm_2) }
        Info: 1 tuple computed. Total elapsed time: 295 ms.

Query 2: Ask for value of non-farm households.
DES-Datalog> /assert query2(X) :- table1(A1,A2,A3,A4,A5,A6,A7,A8,A9,A10,A11,A12,
        A13,A14,A15,A16,A17,A18,A19,A20,A21,A22,A23,A24,A25,
        A26,A27,A28,A29,A30,A31,A32,A33,A34), A4=farm_1, X=A6.
DES-Datalog> query2(X).
        { query2(value_1), query2(value_2), query2(value_3), query2(value_4),
        query2(value_5), query2(value_6), query2(value_7) }
        Info: 7 tuples computed. Total elapsed time: 1029 ms.
Search for semantic rules to transform query: (No rules applied; waste time 296 ms.)

Query 3: Ask for total family income for a household unit with two families.
DES-Datalog> /assert query3(X):-table1(A1,A2,A3,A4,A5,A6,A7,A8,A9,A10,A11,A12,
        A13,A14,A15,A16,A17,A18,A19,A20,A21,A22,A23,A24,A25,
        A26,A27,A28,A29,A30,A31,A32,A33,A34),A24=famunit_2,X=A8.
DES-Datalog> query3(X).
        { query3(ftotinc_1), query3(ftotinc_2), query3(ftotinc_3) }
        Info: 3 tuples computed. Total elapsed time: 906 ms.
Search for semantic rules to transform query: (No rules applied; waste time 282 ms.)

Query 4: Ask for family size of non-farm households with family income at level 3
        (10,000-99,999 US$).
DES-Datalog> /assert query4(X):-table1(A1,A2,A3,A4,A5,A6,A7,A8,A9,A10,A11,A12,
        A13,A14,A15,A16,A17,A18,A19,A20,A21,A22,A23,A24,A25,
        A26,A27,A28,A29,A30,A31,A32,A33,A34),A4=farm_1,A8=ftotinc_3,X=A21.
DES-Datalog> query4(X).
        { query4(famsize_1), query4(famsize_2), query4(famsize_3), query4(famsize_4),
        query4(famsize_5), query4(famsize_6), query4(famsize_7) }
        Info: 7 tuples computed. Total elapsed time: 1028 ms.
Search for semantic rules to transform query:
        DES-Datalog> /assert p(farm_1).
        DES-Datalog> /assert p(ftotinc_3).
        DES-Datalog> p(C).
        { p(farm_1), p(ftotinc_3), p(gq_1), p(gqtypeg_0) }
        Info: 6 tuples computed. Total elapsed time: 594 ms.
Query processing result (semantically transformed query):
        (Time saving = 1028-(594+94)= 340 ms.)
        DES-Datalog> /assert query4B(X):-table1(A1,A2,A3,A4,A5,A6,A7,A8,A9,A10,
        A11,A12,A13,A14,A15,A16,A17,A18,A19,A20,A21,A22,A23,A24,
        A25,A26,A27,A28,A29,A30,A31,A32,A33,A34). A4=farm_1,
        A8=ftotinc_3, A2=gq_1, A3=gqtypeg_0, X=A21.
        DES-Datalog> query4B(X).
        { query4B(famsize_1), ..., query4B(famsize_7) }
        Info: 7 tuples computed. Total elapsed time: 94 ms.

Query 5: Ask for educational record and total income of female person. (Joining of
table1 and table2)

DES-Datalog> /assert query5(X,Y):-table1(A1,A2,A3,A4,A5,A6,A7,A8,A9,A10,A11,
        A12,A13,A14,A15,A16,A17,A18,A19,A20,A21,A22,A23,A24,A25,
        A26,A27,A28,A29,A30,A31,A32,A33,A34),table2(B1,B2,B3,B4,B5,
        B6,B7,B8,B9,B10,B11,B12,B13,B14,B15,B16,B17,B18,B19,B20,
        B21,B22,B23,B24,B25,B26,B27),A1=B1,A30=sex_2,X=B3,Y=B15.
DES-Datalog> query5(X,Y).
    { query5(educrec_0,incwage_4), ..., query5(educrec_9,incwage_4) }
    Info: 756 tuples computed. Total elapsed time: 10423 ms.

Transformed query with materialized view:(Time saving = 10423-1060 = 9363 ms.)
DES-Datalog> /assert query5B(X,Y):-view1(V1,V2,V3,V4,V5,V6,V7,V8,V9,V10,V11,
        V12,V13,V14,V15,V16,V17,V18,V19,V20,V21,V22,V23,V24,V25,
        V26,V27,V28,V29,V30,V31,V32,V33,V34,V35,V36,V37,V38,V39,
        V40,V41,V42,V43,V44,V45,V46,V47,V48,V49,V50,V51,V52,V53,
        V54,V55,V56,V57,V58,V59),V29=sex_2,X=V35,Y=V47.
DES-Datalog> query5B(X,Y).
    { query5B(educrec_0,incwage_4), ..., query5B(educrec_9,incwage_4) }
    Info: 33 tuples computed. Total elapsed time: 1060 ms.

Query 6: Ask for class of work, occupation, and income of white immigrants moving
from Europe and being a single family unit. (Joining table1 and table2)

DES-Datalog> /assert query6(X,Y,Z):-table1(A1,A2,A3,A4,A5,A6,A7,A8,A9,A10,A11,
        A12,A13,A14,A15,A16,A17,A18,A19,A20,A21,A22,A23,A24,A25,
        A26,A27,A28,A29,A30,A31,A32,A33,A34),table2(B1,B2,B3,B4,B5,
        B6,B7,B8,B9,B10,B11,B12,B13,B14,B15,B16,B17,B18,B19,B20,
        B21,B22,B23,B24,B25,B26,B27), A1=B1, A9=nfams_1,
        A31=raceg_1, A34=bplg_4, X=B9, Y=B27, Z=B14.
DES-Datalog> query6 (X,Y,Z).
    { query6 (classwkg_0,occupation_5,inctot_1, ...,
      query6 (classwkg_2,occupation_2,inctot_3) }
    Info: 121 tuples computed. Total elapsed time: 9852 ms.

Transformed query with semantic rules and views:
    DES-Datalog> /assert p(nfams_1)
    DES-Datalog> p(C)
        { p(famunit_1), p(farm_1), p(nfams_1) }
        Info: 3 tuples computed. Total elapsed time: 452 ms.

Query processing result: (Time saving = 9852-(452+1015)= 8385 ms.)
    DES-Datalog> /assert query6B (X,Y,Z):-view1 (V1,V2,V3,V4,V5,V6,V7,V8,V9,V10,
        V11,V12,V13,V14,V15,V16,V17,V18,V19,V20,V21,V22,V23,V24,V2
        5,V26,V27,V28,V29,V30,V31,V32,V33,V34,V35,V36,V37,V38,V39,
        V40,V41,V42,V43,V44,V45,V46,V47,V48,V49,V50,V51,V52,V53,
        V54,V55,V56,V57,V58,V59), V3=farm_1, V8=nfams_1,
        V23=famunit_1, V30=raceg_1, V33=bplg_4, X=V41, Y=V59,
        Z=V46.
    DES-Datalog> query6B (X,Y,Z).
        { query6B (classwkg_0,occupation_5,inctot_1), ...,
          query6B (classwkg_2,occupation_2,inctot_3) }
        Info: 11 tuples computed. Info: Total elapsed time: 1015 ms.

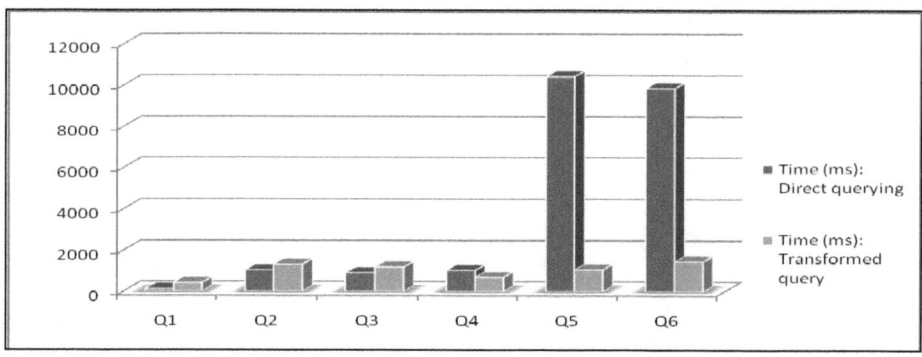

Fig. 7. Time comparison of direct querying versus transforming queries with materialized views and data mining models prior to accessing the database contents

Table 1. Processing time summarization of direct querying versus intelligent querying based on semantic transformation using materialized views and mining models

Query characteristics	Processing time (ms)		Reduced time (ms)	Time usage efficiency
	Direct answer	Intelligent answer		
Query1: ask one information with a single condition, null answer	110	405	-295	-268.18%
Query2: ask one information with a single condition	1,029	1,325	-296	-28.76%
Query3: ask one information with a single condition	906	1,188	-282	-31.12%
Query4: ask one information with two conditions	1,028	688	340	33.07%
Query5: ask two information with a single condition	10,423	1,060	9,363	89.83%
Query6: ask three information with four conditions	9,852	1,467	8,385	85.10%

5   Conclusion

We design and implement a query answering system to provide an integrated and efficient platform for the next generation database management system. To answer queries effectively and intelligently, the association mining component and the materialized view manager are two key players to derive useful knowledge relevant to the given query. Query rewriter, which is supported by intelligent transformation rules and co-operated with query executor, is expected to produce answers in an intelligent way. The preliminary experimental results satisfy the expectation. We are, however, improving the capability of these components to analyze the user's intent and preferences to better providing associated information. Extending the scope of this research towards the distributed environment is also the direction of our future work.

Acknowledgments. This work has been supported by grants from the National Research Council of Thailand (NRCT) and Suranaree University of Technology via the funding of Data Engineering Research Unit.

References

1. Afrati, F.N., Li, C., Ullman, J.D.: Using views to generate efficient evaluation plans for queries. Journal of Computer and System Sciences 73(5), 703–724 (2007)
2. Agrawal, R., Srikant, R.: Fast algorithm for mining association rules. In: VLDB, pp. 487–499 (1994)
3. Arago, M., Fernandes, A.: Logic-based integration of query answering and knowledge discovery. In: 6th Flexible Query Answering Systems, pp. 68–83 (2004)
4. Blockeel, H., Calders, T., Fromont, E., Goethals, B.: Mining views: data base views for data mining. In: IEEE ICDE, pp. 1608–1611 (2008)
5. Calders, T., Goethals, B., Prado, A.: Integrating Pattern Mining in Relational Databases. In: Fürnkranz, J., Scheffer, T., Spiliopoulou, M. (eds.) PKDD 2006. LNCS (LNAI), vol. 4213, pp. 454–461. Springer, Heidelberg (2006)
6. Chang, J., Lee, S.: Query reformulation using materialized views in data warehousing environment. In: ACM Int. Workshop on Data Warehousing and OLAP, pp. 54–59 (1998)
7. Chaudhuri, S.: Generalization and a framework for query modification. In: IEEE ICDE, pp. 138–145 (1990)
8. Chaudhuri, S., Krishnamurthy, S., Potamianos, S., Shim, K.: Optimizing queries with materialized views. In: IEEE ICDE, pp. 190–200 (1995)
9. Chaudhuri, S., Narasayya, V., Sarawagi, S.: Extracting predicates from mining models for efficient query evaluation. ACM Trans. Database Systems 29(3), 508–544 (2004)
10. Chen, C.M., Rossopoulos, N.: The Implementation and Performance Evaluation of the ADMS Query Optimizer: Integrating Query Result Caching and Matching. In: Jarke, M., Bubenko, J., Jeffery, K. (eds.) EDBT 1994. LNCS, vol. 779, pp. 323–336. Springer, Heidelberg (1994)
11. Chu, W., Chen, Q.: A structured approach for cooperative query answering. IEEE Trans. Knowledge and Data Engineering 6, 738–749 (1994)
12. Datalog Educational System, version 2.0, http://www.fdi.ucm.es/profesor/fernan/DES/
13. Erlang Programming Language, release 14, http://www.erlang.org
14. Gou, G., Kormilitsin, M., Chirkova, R.: Query evaluation using overlapping views: completeness and efficiency. In: ACM SIGMOD, pp. 37–48 (2006)
15. Halevy, A.: Answering queries using views: a survey. The VLDB Journal 10(4), 270–294 (2001)
16. Han, J., Huang, Y., Cercone, N., Fu, Y.: Intelligent query answering by knowledge discovery techniques. IEEE Trans. Knowledge and Data Engineering 8(3), 373–390 (1996)
17. IBM, IBM intelligent miner scoring, administration and programming for DB2 version 7.1. IBM, New York (2001)
18. Integrated Public Use Microdata Series (IPUMS), Minnesota Population Center, http://www.ipums.org
19. Lin, T., Cercone, N., Hu, X., Han, J.: Intelligent query answering based on neighborhood systems and data mining techniques. In: IEEE IDEAS, pp. 91–96 (2004)
20. Microsoft Corporation, OLE DB for data mining. Microsoft Corporation, Redmond (2000)
21. Muslea, I.: Machine learning for online query relaxation. In: ACM SIGMOD, pp. 246–255 (2004)
22. Srivastava, D., Das, S., Jagadish, H.V., Levy, A.Y.: Answering queries with aggregation using views. In: VLDB, pp. 318–329 (1996)

IPC Code Analysis of Patent Documents Using Association Rules and Maps – Patent Analysis of Database Technology

Sunghae Jun

Depatment of Bioinformatics and Statistics, Cheongju University,
360764 Chungbuk, Korea
shjun@cju.ac.kr

Abstract. Patent documents are the results of researched and developed technologies. Patent is a protecting system of inventors' right for their technologies by a government. Also, patent is an important intellectual property of a company. R&D strategy has been depended on patent management. For efficient management of patent, we need to analyze patent data. In this paper, we propose a method for analyzing international patent classification (IPC) code as a patent analysis. We introduce association rules and maps for IPC code analysis. To verify our improved the performance, we will make experiments using searched patent documents of database technology.

Keywords: IPC code, Patent analysis, Association rule, Association map, Database technology.

1 Introduction

Most results of researched and developed technologies have been published as patent or paper. These consist of huge literatures with text documents. Using the document data, we can forecast the future trend of a given technology. Patent analysis is one of many approaches for technology forecasting [1-3]. But, it is difficult to analyze the patent documents by quantitative methods such as statistics and machine learning because the methods demand numeric and categorical data as input data [4]. So, most analytical methods of the patent data for technology forecasting were based on some qualitative methods such as Delphi [5-8]. These methods were not stable because they were depended on experts' prior knowledge subjectively. So, we have needed an objective method for analyzing the patent documents. Some approaches for objective analysis of the patent documents were published [9-11]. These works focused on the text analysis from titles and abstracts in patent documents. In this paper, we propose a quantitative method for technology forecasting as an objective method. We use association rules and maps as objective analysis and patent documents as objective data. Association rule method is popular data mining method based on probability [12-14]. Association map is a visualization constructed by extracted association rules. In this paper, we will use patent documents of 'database technology' as a given technology field for technology forecasting. The patent data will be retrieved from

USPTO (United State Patent and Trademark Office, www.uspto.gov). A patent document has many descriptions of applied technology such as applied date, inventor, patent number, abstract, title, claims, drawing, citation, international patent classification (IPC) code, and so on. Among these descriptions, we select IPC codes as input data for patent analysis. Also, we will forecast vacant technology of a given technology field. We determine the vacant technology areas using the results of association rules and maps. In next section, we will review patent system and IPC code. We will introduce proposed method for forecast vacant technology in section 3. To verify the performance our work, we will show experimental results in section 4. Final section will include our conclusion and future work.

2   Patent and International Patent Classification Code

Patent is a protection system of inventors' right of their developed technology for a certain period of time. The right of issued patent is protected by national governments in the world. Patent is a document with technological information such as patent number, issued date, inventor, assignee, abstract, title, IPC code and so on. That is, a patent document is large literature including technological information. We knew that company's success and its patent management was highly associated from the previous researches [15-17]. Patent analysis is an approach for efficient management of the patent. Most methods of patent analysis were based on text mining and citation analysis [18-22]. In this paper, we propose another approach to patent analysis. We introduce IPC code analysis as a method of patent analysis. We do not use the results of text mining such as keyword extraction from the titles and abstracts. Our research focuses on the associations between IPC codes. The IPC is a hierarchical patent classification established by the Strasbourg Agreement 1971 [23]. It has been administered by World Intellectual Property Organization (WIPO). First level of IPC code consists of a letter from A to H. They are Human Necessities (A), Performing Operations, Transporting (B), Chemistry, Metallurgy (C), Textiles, Paper (D), Fixed Constructions (E), Mechanical Engineering, Lighting, Heating, Weapons (F), Physics (G), and Electricity (H). The subclass is then followed by main group of IPC code. Next figure shows an example of hierarchical IPC code [23].

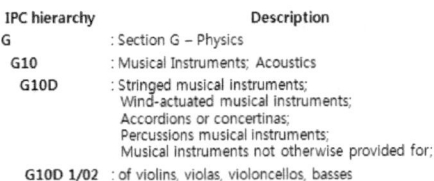

Fig. 1. Hierarchical structure of IPC codes

IPC code represents a technological classification term indicating the invented subject. In this paper, we use the 3rd level of IPC code such as G10D in Fig. 1. Most IPC code hierarchy of patent documents is based on this code level.

3   Association Rules and Maps for IPC Code Analysis

In this paper, we propose a method of IPC code analysis using association rules and maps. IPC code data of patent documents are so large. Mining association rules is an effective approach to extract meaningful relationships between items in large data [12],[22],[24]. Data set for association rules consists of item and transaction sets. $I=\{i_1, i_2, ..., i_n\}$ is an item set and $T=\{t_1, t_2, ..., t_m\}$ is a transaction set. A t_i of transaction set has a unique number and contains items [3]. When X and Y are items, $(X \rightarrow Y)$ is defined as an association rule. X and Y represent antecedent and consequent of an association rule respectively. In our research, transaction and item are patent document and its IPC code respectively. For example, we have an item set, $I=\{G10D, A06F, H04C, B07D\}$ and a database of transactions is shown as follow.

Table 1. Transactions and items in patent document data

Number of patent documents (transactions)	Terms (items)
US09998219	G10D, A06F
US12472680	A06F, H04C
US08918106	B07D
US12719217	G10D, A06F, H04C
US12157562	A06F, H04C

A association $(A06F \rightarrow G10D)$ means that if $A06F$ technology is developed, $G10D$ technology will be developed. This can be more detailed as follow [24].

develop technology (X, "A06F") → develop technology (Y, "G10D")

We need three measures to extract meaningful rules from the set of all possible rules. The measures are support, confidence, and lift. We select association rules with minimum cut off values on support and confidence. So, we find the meaningful rules with large lift from the result by support and confidence. The support of items X and Y is defined as follow.

$$S(X \rightarrow Y) = P(X \cap Y) \tag{1}$$

That is, the support is the proportion of transactions of a rule in all observed rules. In the example, the support of $(A06F \rightarrow G10D)$ is computed as follow.

$$S(A06F \rightarrow G10D)=P(A06F \rightarrow G10D)=2/5=0.4 \tag{2}$$

It occurred in 40% of all transactions. Actually, in all technological developments, the technologies including A06F and G10D are developed in 40%. The confidence of an association rule is defined as following probability.

$$C(X \rightarrow Y) = P(Y \mid X) = \frac{P(X \cap Y)}{P(X)} = \frac{S(X \rightarrow Y)}{S(X)} \qquad (3)$$

The confidence value is equal to the constraint support value in not all transactions but the transactions with occurred X. For example, the confidence of $(A06F \rightarrow H04C)$ is computed as follow.

$$C(A06F \rightarrow H04C) = \frac{P(A06F \cap H04C)}{P(A06F)} \qquad (4)$$
$$= (3/5)/(4/5) = 0.75$$

We have to determine the thresholds of support and confidence to extract the meaningful association rules. In general, many rules are extracted from transaction database. It is difficult to select the meaningful rules in the extracted rules. Another measure to solve this problem is lift. This is defined as follow.

$$L(X \rightarrow Y) = \frac{P(Y \mid X)}{P(Y)} = \frac{P(X \cap Y)}{P(X)P(Y)} = \frac{S(X \rightarrow Y)}{S(X)S(Y)} \qquad (5)$$

The support and confidence measures are probability. But, lift measure is not probability. The values of lift measure are from 0 to ∞. If the value is 1, X and Y are independent each other. That is, they are absolutely not associated. Larger lift values demonstrate stronger associations. In the table 1, the lift value of $(H04C \rightarrow A06F)$ is computed as follow.

$$L(H04C \rightarrow A06F) = \frac{P(H04C \cap A06F)}{P(H04C)P(A06F)}$$
$$= \frac{3/5}{(3/5)(4/5)} = 1.25 \qquad (6)$$

Increasing the development of technology $H04C$ bring the development of technology $A06F$. If the lift value of technology X and technology Y is below 1, X and Y are substitutive each other. That is, technology X is developed, then, technology Y is not needed to be developed. In this paper, we use scatter plot and association map for visualization of the meaningful rules. In previous research, Bayardo and Agrawal used support, confidence, and lift as axes of scatter plot [25]. In the association map, the diameter of a circle represents the size of support and the intensity of the color shows the size of confidence. So, we can find meaningful association rules using scatter plot and association map. We account for our method step-by-step as follow. In this paper, we determine 'database theory and application (DTA)' as a given technology field (database technology) for vacant technology forecasting.

Vacant Technology Forecasting of Database Theory and Application (DTA)
Step1. Association rules
(1-1) Constructing keywords equation for DTA;
(1-2) Retrieving patent documents of DTA from USPTO;
(1-3) Selecting IPC codes in the patent documents;
(1-4) Extracting association rules from IPC code data
     by support, confidence, and lift;
(1-5) Finding vacant technology areas in extracted
     rules;
Step2. Association maps
(2-1) Constructing scatter plot with support,
     confidence, and lift on the axes;
(2-2) searching vacant technology areas in vacant or
     relatively sparse areas in the scatter plot;
(2-3) Constructing association map;
(2-4) Finding vacant technology areas from
visualization of association map;
Step3. Determination of vacant technology for DTA
(3-1) Combining the results of association rules and
     maps;
(3-2) Determining final areas for vacant technology
     forecasting of DTA;

Though we selected 'database theory and application' as a given technology field, our method can be applied to any technology field for technology forecasting. Using the results of association rules and maps, we determine the vacant areas of technology forecasting. Next, we make some experiments to verify the performance of our work using patent documents of DTA. Though, we select DTA as a given technology field in this paper, our proposed model can be applied to all technology fields.

4 Experimental Results

For vacant technology forecasting by IPC code analysis, we retrieved patent documents from USPTO using the following keyword equation.

Title = databases * (multimedia + spatial + active + biomedical + classification + ranking + clustering + mining + visualization + integration + management + ubiquitous + mobile + extraction + modeling + provenance + quality + security + privacy + streaming + heterogeneous + warehousing + OLAP + parallel + distributed + query + processing + reliability + semantic + metadata + analysis + temporal + services)

In this paper, we tried to retrieve the patent documents about the technologies of DTA. We applied the keywords equation to the title of patent document. In the equation, '+' and '*' are 'or' and 'and' operators, respectively. We retrieved the patent documents until July 11, 2011. Finally, we got 3983 patent documents after removing the patents not related to DTA. In this paper, we used R-project packages for patent analysis [26-28]. The first patent about DTA was applied in 1983. The number of applied patents had been on the increase at the mid 1990s. The rate of increase had been fast at the first 2000s. Using all retrieved patent documents, we constructed transaction data with 206 transactions (rows) and 46 items (columns). Following table shows the top ten IPC codes and their explanations.

Table 2. Top ten IPC codes and technological explanations

Top ten IPC codes	Code technologies
1. G06F	Electric digital data processing
2. H04L	Transmission of digital information, telegraphic communication
3. H04M	Telephonic communication
4. H04Q	Selecting
5. G06Q	Data processing systems or methods
6. G01C	Measuring distances, levels or bearings, surveying, navigation
7. G06K	Recognition and presentation of data, record carriers
8. G01N	Investigating or analyzing materials
9. H04N	Pictorial communication, e.g. television
10. G06N	Computer systems based on specific computational models

We knew the technologies of 'electric digital data processing (*G06F*)' and 'transmission of digital information, telegraphic communication (*H04L*)' were occurred in the most patents of DTA. This result shows the support of one item, *S(X)*. Next, we searched the meaningful rules of items *X* and *Y*. The following two scatter plots show the distribution of IPC codes by support, confidence, and lift.

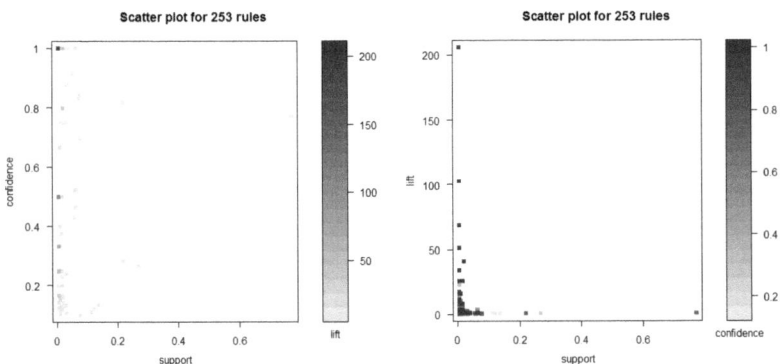

Fig. 2. Scatter plot of association rules

Left figure is the scatter plot with support and confidence on the axes. The lift values are increased according as the colors of points are darker. The scatter plot of right figure uses support and lift on the axes. Also, the darker the color of a point is, the larger the confidence value is. From these figures, we found the candidate areas of vacant technology in small support, large confidence, and large lift values. These were vacant and relatively sparse areas. We forecasted the vacant technology of DTA broadly using the scatter plots of association rules. Next, we extracted the meaningful association rules by support, confidence, and lift values.

Table 3. Top three rules generated by support

X→Y	Rank	Support	Confidence	Lift
H04L→G06F G06F→H04L	1	0.2184	0.8182 0.2830	1.0600
G01C→G06F G06F→G01C	2	0.0777	0.8421 0.1006	1.0910
G06Q→G06F G06F→G06Q	3	0.0777	0.6957 0.1006	0.9013

We knew two codes of the technology pairs, (*H04L*, *G06F*), (*G01C*, *G06F*), and (*G06Q*, *G06F*) were associated each other. That is, if technology H04L was developed, technology G06F was also developed. But, from the results of scatter plot, we found the candidate areas of vacant technology had small support values. So, we focus on the results of large confidence and lift measures.

Table 4. Top three rules generated by confidence

X→Y	Rank	Confidence	Support	Lift
A61N→A61B	1	1	0.0049	34.3333
G08G→G06F	2	1	0.0049	1.2956
G02C→A61B	3	1	0.0049	34.3333

All rules were large confidence and small support. So, they were satisfied with the conditions of vacant technology by support and confidence measures. Another condition of vacant technology was large lift. The rule, (*G08G→G06F*) was not a candidate of vacant technology because its lift value was 1.2956. It was so small. So, we decided two rules, (*A61N→A61B*) and (*G02C→A61B*) were meaningful. Next, we forecast the vacant technology of DTA using lift values.

Table 5. Top ranked rules by lift

Measure values	Generated rules
Lift=206 Support=0.0049 Confidence=1	G11C→G11B {G01C, G01S}→B60K {G01C, H04Q}→B60K {G01R, H01L}→H03K {G01C, G01S, H04Q}→B60K {G01R, G06F, H01L}→H03K

From the result of scatter plot, we knew that the vacant technology areas had small support, large confidence, and large lift. Table 5 shows the rules satisfied with the conditions of vacant technology from the scatter plot. So, we got six rules in Table 5.

These were considered as strong candidates for vacant technology areas of DTA. In this paper, we proposed the final vacant technology areas were determined from the results of association rules and maps. Next, the association map for vacant technology forecasting of DTA is shown.

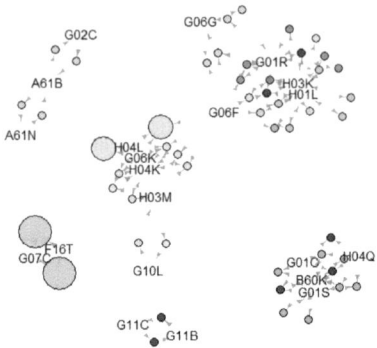

Fig. 3. Association map for vacant technology forecasting of DTA

The diameter of a circle represents the size of support. Also, the intensity of the color shows the size of confidence. So, we selected the areas with small and dark circles for vacant technology forecasting. Using the results of Table 5 and Fig. 6, we decided the vacant technology areas of DTA as follow.

Table 6. Final selected rules for vacant technology forecasting of DTA

Generated rules	IPC technologies
G11C→G11B {G01C, G01S}→B60K {G01C, H04Q}→B60K {G01R, H01L}→H03K {G01R, G06F, H01L}→H03K	**G11C** – Static stores **G11B** – Information storage based on relative movement between record carrier and transducer **G01C** – Measuring distances, levels or bearings, surveying, navigation **G01S** – Radio direction-finding; Radio navigation … **B60K** – Arrangement or mounting of propulsion units or of transmissions in vehicles **H04Q** – Selecting **G01R** – Measuring electric variables; measuring magnetic variables **H01L** – Semiconductor devices; Electric solid state device not otherwise provided for **H03K** – Pulse techniques **G06F** – Electric digital data processing

We found five rules in Table 6. For example, in the first rule (G11C→G11B), if technology G11C is developed, technology G11B will also be developed. That is, the technology of 'static stores' is developed, the technology of 'Information storage based on relative movement between record carrier and transducer' will also be developed. This rule, (G11C→G11B) will be needed for technological development of DTA in the future. In addition, the rules ({G01C, G01S}→B60K), ({G01C,

H04Q}→B60K), ({G01R, H01L}→H03K), and ({G01R, G06F, H01L}→H03K) were selected as vacant technologies of DTA in our work.

5   Conclusions and Future Works

In this paper, we proposed a model of vacant technology forecasting of DTA using association rules and maps. For the experiments to verify the performance of our work, we used the retrieved patent documents of DTA from USPTO. We only used IPC codes in the patent documents in this paper. Using support, confidence, and lift measures, we searched the vacant technology area in extracted association rules. Also, we found the vacant technology areas in the scatter plot and association map. Finally we combined the results of association rules and maps for forecasting the vacant technology of DTA. In this paper, we tried to introduce an objective approach for technology forecasting. We used association rules and maps as objective analysis and patent documents as objective data. A limitation of our work was a shortage of DTA domain experts' prior knowledge. So, we suggest to combine our result and given technological knowledge of DTA to get more accurate forecasting of the vacant technology. In our future works, we will develop more advanced method for technology forecasting using combining association rules and other data mining techniques.

References

1. Zhu, D., Porter, A.L.: Automated extraction and visualization of information for technological intelligence and forecasting. Technological Forecastingand Social Change 69, 495–506 (2002)
2. Coates, V., Farooque, M., Klavans, R., Lapid, K., Linstone, H.A., Pistorius, C., Porter, A.L.: On the future of technological forecasting. Technological Forecasting and Social Change 67, 1–17 (2001)
3. Mann, D.L.: Better technology forecasting using systemic innovation methods. Technological Forecasting and Social Change 70, 779–795 (2003)
4. Tseng, Y.H., Lin, C.J., Lin, Y.I.: Text mining techniques for patent analysis. Information Processing & Management 43, 1216–1247 (2007)
5. Madu, C.N., Kuei, C.H., Madu, A.N.: Setting priorities for IT industry in Taiwan-A Delphi study. Long Range Planning 24(5), 105–118 (1991)
6. Mitchell, V.W.: Using Delphi to Forecast in New Technology Industries. Marketing Intelligence & Planning 10(2), 4–9 (1992)
7. Woundenberg, F.: An evaluation of Delphi. Technological Forecasting and Social Change 40, 131–150 (1991)
8. Yun, Y.C., Jeong, G.H., Kim, S.H.: A Delphi technology forecasting approach using a semi-Markov concept. Technological Forecasting and Social Change 40, 273–287 (1991)
9. Yoon, B., Park, Y.: Development of New Technology Forecasting Algorithm: Hybrid Approach for Morphology Analysis and Conjoint Analysis of Patent Information. IEEE Transactions on Engineering Management 54(3), 588–599 (2007)
10. Jun, S., Park, S., Jang, D.: Forecasting Vacant Technology of Patent Analysis System using Self Organizing Map and Matrix Analysis. Journal of the Korea Contents Association 10(2), 462–480 (2010)

11. Jun, S., Uhm, D.: Patent and Statistics, What's the connection? Communications of the Korea Statistical Society 17(2), 205–222 (2010)
12. Hahsler, M., Grun, B., Hornik, K.: arules – A Computational Environment for Mining Association Rules and Frequent Item Sets. Journal of Statistical Software 14(15), 1–25 (2005)
13. Agrawal, R., Imielinski, T., Swami, A.: Mining Association Rules between Sets of Items in Large Databases. In: Proceedings of the 1993 ACM SIGMOD International Conference on Management of Data, pp. 207–216 (1993)
14. Agrawal, R., Mannila, H., Srikant, R., Toivonen, H., Verkamo, A.I.: Fast discovery of association rules. In: Advances in Knowledge Discovery and Data Mining. AAAI/MIT Press (1995)
15. Lerner, J.: The importance of patent scope: an empirical analysis. RAND Journal of Economics 25, 319–332 (1994)
16. Ernst, H.: Patent applications and subsequent changes of performance: evidence from time-series cross-section analyses on the firm level. Research Policy 30, 143–157 (2001)
17. Shane, S.: Technological opportunities and new firm creation. Management Science 7, 205–220 (2001)
18. Nizar, G., Khaled, K., Rose, D.: Supporting Patent Mining by using Ontology-based Semantic Annotations. In: Proceedings of International Conference on Web Intelligence, pp. 435–438 (2007)
19. Wu, C., Ken, Y., Huang, T.: The Support Vector Machine Classification System for Patent Document Information Importance Analysis. In: Proceedings of International Conference on BioMedical Engineering and Informatics, pp. 375–379 (2008)
20. Yoon, B., Park, Y.: Development of New Technology Forecasting Algorithm: Hybrid Approach for Morphology Analysis and Conjoint Analysis of Patent Information. IEEE Transactions on Engineering Management 54(3), 588–599 (2007)
21. Yoon, B., Lee, S.: Patent analysis for technology forecasting: Sector-specific applications. In: Proceedings of IEEE International Conference on Engineering Management, pp. 1–5 (2008)
22. Brinn, M.W., Fleming, J.M., Hannaka, F.M., Thomas, C.B., Beling, P.A.: Investigation of forward citation count as a patent analysis method. In: Proceedings of Systems and Information Engineering Design Symposium, pp. 1–6 (2003)
23. International Patent Classification (IPC), World Intellectual Property Organization (WIPO), http://www.wipo.int/classifications/ipc/en/
24. Han, J., Kamber, M.: Data Mining Concepts and Techniques. Morgan Kaufmann (2001)
25. Bayardo, Jr, R. J., Agrawal, R.: Mining the most interesting rules. In: KDD 1999: Proceedings of the fifth ACM SIGKDD International Conference on Knowledge Discovery and Data Mining, pp. 145–154 (1999)
26. R Development Core Team.: R, A language and environment for statistical computing. R Foundation for Statistical Computing (2011), http://www.R-project.org
27. Hahsler, M., Buchta, C., Gruen, B., Hornik, K.: Package 'arules'. R-project CRAN (2011)
28. Hahsler, M., Chelluboina, S.: Package 'arulesViz'. R-project CRAN (2011)

CAIG: Classification Based on Attribute-Value Pair Integrate Gain

Tianzhong He, Zhongmei Zhou, Zaixiang Huang, and Xuejun Wang

Lab of Granular Computing, Zhangzhou Normal University,
Zhangzhou 363000, China
hetianzhong@163.com

Abstract. Many studies have shown that rule-based classifiers perform well in classifying categorical data. However, a limitation of many of them lies in the use of only one measure to select the best attribute-value pair. In this way, many attribute-value pairs have the same best values. Consequently, it is difficulty to distinguish which attribute-value pair is the best.On the other hand, these classifiers usually combine two best attribute-value pairs to generate rules, whether they bias toward the same class label or not. In this paper, we propose a new classification approach named CAIG (Classification based on Attribute-value pair Integrate Gain) which has a number of new features. First, it uses multi measures to select the best attribute-value pairs. Second, it divides attribute-value pairs into different groups according to the class label. Third, it arranges attribute-value pairs to the same groups if they bias to same class label. Fourth, it adopts a greedy algorithm to generate rules from the theses groups. Experimental results show that the method of multi measures is highly accuracy in comparison with the one measure.

Keywords: Classification, Entropy, Multi measure, Deviation, Effect.

1 Introduction

As one of the most fundamental data mining tasks, classification has been extensively studied. Different classification approaches have been proposed, such as decision tress [1], rough set approaches [2–10], associative classification [11–13], KNN [14–18] and etc.. Among them, one category is the rule-based classifiers [1, 19–21]. They build a model from the training database as a set of high-quality rules, which can be used to predict the class labels of unlabeled instances. Many studies have shown that rule-based classification algorithms perform very well in classifying categorical databases [20, 13, 11, 22]. However, these algorithms may suffer from two major deficiencies.

First, many rule-based algorithms like C4.5 [1], FOIL [20], and PRM [19] discover a set of classification rules using a single measure to select a best attribute-value pair. In many cases, attribute-value pairs have the same best value according a single measure. For example, C4.5 only uses the attribute information gain to select a best attribute-value pair. PRM only uses Foil gain

T.-h. Kim et al. (Eds.): DTA/BSBT 2011, CCIS 258, pp. 31–40, 2011.
© Springer-Verlag Berlin Heidelberg 2011

to select a best attribute-value pair. Unfortunately, it is inappropriate to select a best attribute-value pair according to a single measure. In other words, with single measure, the best attribute-value pair can not be selected correctly.

Second, these rule-based algorithms induce rules directly from all attribute-value pair space. FOIL, and PRM repeatedly searches for the current best rule and removes all the positive instances covered by the rule until all the positive instances in the data set are covered. The main drawback of this approach is that the attribute-value pairs of a rule do not bias toward the same class label .

In this paper, we proposed a new approach, called CAIG, to address these issues. First, CAIG uses a new multi measure method to select the best attribute-value pair, which integrates attribute-values pair entropy and attribute-value pair deviations to select the best attribute-value pair. In comparison with single measure, multi measures can greatly decrease the number of attribute-value pair with same best values. Second, CAIG divides all attribute-value pairs into positive group and the negative group on basis of biased toward class label. CAIG generates rules from the positive attribute-value pair group and the negative attribute-value pair group respectively. Third, when an uncertain class label example satisfies a set of rules, CAIG adopts rule's strength to classify it. Our experimental results show that the techniques developed in this paper is better performance than one measure.

The rest of the paper is organized as follows. Section 2 introduces some basic definitions and notations. Section 3, we describe classification based on Attribute-value pair Integrate Gain. The experimental results are presented in section 4, and we conclude the study in section 5.

2   Rule Based Classication

A set of tuples T has m distinct attributes $A_1, A_2, \ldots, A_m$ and class $C = \{C_1, C_2\}$. Each tuple $\{a_1, a_2, \ldots, a_m, c\}$, called instance, where a_i is the value of the A_i and c is the value of C . A instance $\{a_1, a_2, \ldots, a_m, c_1\}$,called positive instance and a instance $\{a_1, a_2, \ldots, a_m, c_2\}$,called negative instance.

A rule r, which takes the form of $(A_1, a_1) \wedge (A_2, a_2) \wedge \ldots \wedge (A_m, a_m) \Rightarrow c_k, k = \{1, 2\}$,consists of a conjunction of attribute-value pairs$(A_1, a_1), (A_2, a_2) \ldots (A_m, a_m)$, associated with a class label c_k. A instance i satisfies rule r's body if and only if it satisfies every attribute-value pair in the rule. If instance i satisfies rule r's body, and i is of class c_k, called instance i matches rule r.

The entropy of an attribute-value pair (A_i, a_i) is defined by

$$H(A_i, a_i) = -\frac{m_1}{n} \log_2 \frac{m_1}{n} - \frac{m_2}{n} \log_2 \frac{m_2}{n}, \tag{1}$$

where n is the number of the instances which satisfy the attribute-value pair (A_i, a_i). m_1 is the number of positive instances which satisfy attribute-value pair (A_i, a_i). m_2 is the number of negative instances which satisfy attribute-value pair (A_i, a_i).

The deviation of an attribute-value pair (A_i, a_i) is defined by

$$deviation(A_i, a_i) = \frac{2(m_1 - m_2)}{n},\tag{2}$$

where n is the number of the all instances. m_1 is the number of positive instances which satisfy attribute-value pair (A_i, a_i). m_2 is the number of negative instances which satisfy attribute-value pair (A_i, a_i).

The effect of an attribute-value pair(A_i, a_i) is defined by

$$effect(A_i, a_i) = (1 - entropy(A_i, a_i) + |deviation(A_i, a_i)|).\tag{3}$$

In comparison with attribute-value pair entropy, attribute-value pair effect integrates attribute-value pair entropy, attribute-value pair deviation.

The confidence of a rule r is defined by

$$confidence(r) = \frac{m_c}{m},\tag{4}$$

where m_c is the number of the instances which satisfy the rule r'body. m is the number of the instances which match the rule r .

Rule-based algorithms work in three phases:

First, rule-based algorithms employ a function for evaluating attribute-value pairs. Various evaluation criteria have been used in different learning algorithms. For example, C4.5 [1] employs an entropy-based information gain to find the most relevant attribute to grow decision trees. PRISM [22] uses another form of information gain which can be characterized in terms of apparent classification accuracies on the training set to measure the relevance of attribute-value pairs with respect to a target concept. They all only use single measure to find the relevant of attribute-value pairs which leads to suffer from selecting the best attribute-value pair. We use a multi measure method to select the best attribute-value pair, which integrates attribute-values pair entropy and attribute-value pair deviations.

Second, after evaluating attributes or attribute-value pairs. these rule-based algorithms induce rules directly from all attribute-value pairs space. Obviously, it is not appropriate. Instead of searching from all attribute-value pairs space, we divide all attribute-value pairs into positive group and negative group, and generate rules from two groups respectively.

Third, classifying new objects. Some algorithms use the first matching rule [11] and others use multiple rules [13] to predict a new objects' class label. There are different ways to combine the rules that could classify a new object. Some algorithms average the confidences for each category, while others compute a weighted chi-square [13] for each category.

We use the rule-strength to measure a rule,which integrates the laplace expected error estimate [23] and support of the rule, and average the rule-strengths for each category. From our experiments, we achieved high classification accuracy.

3   CAIG: Classification Based on Attribute-Value Pair Integrate Gain

In this section, we develop a new rule-based classification method, called Classification based on Attribute-value pair Integrate Gain (CAIG), which consists of three phases:(1) Selecting attribute-value pair. (2) Inducing rules. (3) Classification using rules. Section 3.1 discusses selecting attribute-value pair. Section 3.2 discusses inducing rules. Section 3.3 discusses classification.

3.1   Selecting Attribute-Value Pair

CAIG induces rules from the positive attribute-value pair group and negative attribute-value pair group respectively. To select attribute-value pair, CAIG conducts the following.

1. Compute the effect for each attribute-value pair.
2. Divide all attribute-value pairs into positive group and negative group.
3. Select the best attribute-value.

The general idea of selecting an attribute-value pair is shown in the following example. Consider the data set as shown in Table 1. The data set is the subset of mushrooms. The last attribute is class. Compute entropy, deviation and effect for every attribute-value pair as shown in Table 2.

Table 1. A training data set

	A	B	C	D	W
x_1	32	55	80	83	90
x_2	33	52	80	85	89
x_3	33	52	80	85	89
x_4	33	55	79	82	90
x_5	34	55	79	82	89
x_6	34	55	77	82	89
x_7	32	55	80	88	90
x_8	33	55	79	82	90

From the Table 2, we can know that some attribute-value pairs can not distinguish them only with one measure, such as the pairs $(A, 32)$ and $(B, 55)$ can not distinguish by their deviation. But we can distinguish them by their effect. Obviously, it is appropriate to select a best attribute-value pair according to a multi measure.

After computing the effect, CAIG deletes the attribute-value who's deviation, effect is zero and the attibute-value who' entropy is one. Then, CAIG divides these attribute-value pairs into tow groups on basis of biased toward class label. In each group, Attribute-value pairs are positive or negative. The two groups are shown in Tables 3 and 4.

Table 2. Attribute-value pair deviation, entropy and effect

Attribute	Value	Deviation	Entropy	Effect
A	32	0.5	0	1.5
A	33	0	1	0
A	34	-0.5	0	1.5
B	55	0.5	0.92	0.58
B	52	-0.5	0	1.5
C	80	0	1	0
C	79	0.25	0.92	0.33
C	77	-0.25	0	1.25
D	83	0.25	0	1.25
D	85	-0.5	0	1.5
D	82	0	1	0
D	88	0.25	0	1.25

Table 3. Positive group

Attribute	Value	Deviation	Entropy	Effect
A	32	0.5	0	1.5
B	55	0.5	0.92	0.58
C	79	0.25	0.92	0.33
D	83	0.25	0	1.25
D	88	0.25	0	1.25

Table 4. Negative group

Attribute	Value	Deviation	Entropy	Effect
A	34	-0.5	0	1.5
B	52	-0.5	0	1.5
C	77	-0.25	0	1.25
D	85	-0.5	0	1.5

3.2   Inducing Rules

After slipting all attribute-value pairs into two group, CAIG induces rules from two group. The inducing rules algorithm based on CAIG is presented in Algorithm 1.

In this example, $min-Confidence$ is set to 0.8, $min-effect$ is set to 0.02, δ is set to 0.1, α is set to 0.2, β is set to 0.01,

CAIG selects the best attribute-value pair and changes it'effect iteratively, until the confidence of the rule achieves the min-confidence. From the Table 3, CAIG selects the attribute-value pairs $(A, 32)$, $(D, 83)$ and $(D, 88)$ and their confidence are 1. CAIG induces three rules as follow:

$r_1 : A = 32 \Rightarrow W = 90 \quad r_2 : D = 32 \Rightarrow W = 90 \quad r_3 : D = 88 \Rightarrow W = 90$

After inducing three rules, the Table 3 is changed as shown Table 5 and current-weight is 2.51

Algorithm 1. CAIG

Input: Two group of the attribute-value pairs.
Output: A rule set for predicting class labels for instances.

1: Initialize Options;
2: R ← ∅;
3: **for all** g in groups **do**
4:     set the weight of every instance to 1;
5:     init-weigth ← current-weight();
6:     **while** $(current - weight() > \alpha * init - weight)$ **do**
7:         fromAttribute ← all attribute;
8:         r ← ∅;
9:         **while** fromAttribute ≠ ∅ **do**
10:             attribute-value $p = \max(g, fromAttribute)$;
11:             **if** $p.effect < min - effect$ **then**
12:                 continue FOR;
13:             **end if**
14:             append p to r;
15:             delete the attribute of the p from fromAttribute;
16:             count the rule r'confidence r_{conf}
17:             **if** $(r_{conf} \geq$ min-confidence$)$ **then**
18:                 **for all** instances ins satisfying r **do**
19:                     $ins.weight = \delta * ins.weight$;
20:                 **end for**
21:             **end if**
22:             $p.effect = \beta * p.effect$;
23:             R ← R∪ r;
24:         **end while**
25:     **end while**
26: **end for**
27: return R;

Table 5. Positive group after inducing three rules

Attribute	Value	Deviation	Entropy	Effect
A	32	0.5	0	0.015
B	55	0.5	0.92	0.58
C	79	0.25	0.92	0.33
D	83	0.25	0	0.0125
D	88	0.25	0	0.0125

Because current-weight is greater than 1.6, inducing rules continues. CAIG selects the attribute-value pairs $(B, 55)$ and $(C, 79)$. The rule confidence is 0.67 and less than min-confidence. After the attribute-value pair $(C, 79)$ selected, the max effect of the positive group is 0.015 and less than min-effect. So, Inducing rules for positive group stop. To learn rules for negative group, the process is repeated. Positive group after the attribute-value pair $(C, 79)$ selected as shown Table 6.

Table 6. positive group after the attribute-value pair $(C, 79)$ selected

Attribute	Value	Deviation	Entropy	Effect
A	32	0.5	0	0.015
B	55	0.5	0.92	0.0058
C	79	0.25	0.92	0.0033
D	83	0.25	0	0.0125
D	88	0.25	0	0.0125

3.3 Classification Using Rules

In this section, we discuss how to determine the class label of new objects.

CAIG collects the subset of rules matching the new object from the set of rules. If all the rules matching the new object have the same class label, CAIG just simply assigns that label to the new object.

If the rules are not consistent in class labels, CAIG divides the rules into two groups according to class labels. All rules in a group share the same class label and each group has a distinct label.

Firstly, we use the Laplace expected error estimate [23] and the support of rule to estimate the accuracy of rule, called rule-strength. the Laplace expected error estimate of a rule r is defined as follows:

$$laplace - accuracy(r) = \frac{n_c + 1}{n_{tot} + k}, \tag{5}$$

where k is the number of classes, n_{tot} is the total number of examples satisfying the rule's body, among which n_c instances belong to c, the predicted class of the rule; The rule-strength of a rule r is defined as follows:

$$rule - strength(r) = laplace - accuracy(r) \times n_c, \tag{6}$$

where n_c is the number of instances which satisfy rule r.

Secondly, we measure the combined effect of a group rule $R\{r_1, r_2, \ldots r_n\}$ by compute its avage strength , called $avage - strength(R)$. It is defined as follows:

$$avage - strength(R) = \frac{\sum_{k=1}^{n} rule - strength(r_k)}{n}, \tag{7}$$

where n is the r's number of the R.

Finally, CAIG assigns the class label of the group with maximum avage-strength to the new object.

4   Experiments

We tested our algorithm on the Mushroom data set. All the experiments are performed on a $2.67GHz$ PC with $1G$ main memory, running Microsoft Windows XP.

To evaluate the accuracy of CAIG, we designed three groups of experiments. There are performed with attribute-value pair entropy measure, with attribute-value pair deviation measure and with the measure of both . In each group experiment, we let the train data set grows from 100 to 1000 with the step 100. One train set and ten test sets are used for each step. The results are given as the average of accuracy, number of rules and run time. All reports of the runtime only include the runtime in rule generation. In the rule generation algorithm, $max - entropy$ is set to 1, $min - deviation$ is set to 0, $min - Confidence$ is set to 0.8, $min - effect$ is set to 0.01, δ is set to 0.01, α is set to 0.6, β is set to 0.18. The results as shown Tables 7, 8, 9. As can be seen from the Tables 7, 8, 9, the accuracy of the multi measure with entropy and deviation is higher than single measure with entropy or deviation.

Table 7. Entropy measure

size	accuracy	rules number	time
100	0.9766	45	0.0465
200	0.9855	45.5	0.0395
300	0.9873	47.0	0.0183
400	0.9892	46.5	0.0626
500	0.9905	46.4	0.0658
600	0.9919	45.1	0.0935
700	0.9926	44.1	0.0445
800	0.9928	44.2	0.0604
900	0.9928	44.0	0.1890
1000	0.9929	44.5	0.2703

Table 8. Deviation measure

size	accuracy	rules number	time
100	0.9950	37.5	0.0075
200	0.9925	36.5	0.0235
300	0.9878	39.5	0.0155
400	0.9892	41.0	0.0841
500	0.9900	40.7	0.0986
600	0.9907	42.1	0.1368
700	0.9915	42.6	0.0725
800	0.9916	43.0	0.0926
900	0.9918	43.6	0.2935
1000	0.9919	43.3	0.3500

5   Conclusions

Rule-based classification algorithms perform well in classifying categorical data. However, many rule-based classifiers suffer from selecting the best attribute-value

Table 9. Entropy and deviation

size	accuracy	rules number	time
100	0.9950	45.0	0.0080
200	0.9950	48.0	0.0275
300	0.9948	46.0	0.0207
400	0.9948	46.6	0.0683
500	0.9944	46.5	0.1219
600	0.9941	46.8	0.1888
700	0.9945	45.8	0.0693
800	0.9944	43.0	0.0927
900	0.9943	44.2	0.2827
1000	0.9942	43.7	0.3110

pair. So,we propose a new classification approach, CAIG (Classification based on Attribute-value pair Integrate Gain). This method has two major features: (1) it uses a new multi measure method to select the best attribute-value pair. In comparison with single measure, multi measures can greatly decrease the number of attribute-value pair with same best values, (2) it divides all attribute-value pairs into positive group and the negative group on basis of biased toward class label, and generates rules from each group respectively. Our experiments on the Mushroom data set show that CAIG achieves higher classification accuracy than single measure.

Acknowledgement. This work is funded by China NSF program (No. 61170129), a grant from education ministry of Fujian, China (No. JA10202).

References

[1] Quinlan, J.R.: C4.5: Programs for machine learning. Morgan Kaufmann (1993)
[2] Zhu, W., Wang, F.: Reduction and axiomization of covering generalized rough sets. Information Sciences 152(1), 217–230 (2003)
[3] Zhu, W.: Generalized rough sets based on relations. Information Sciences 177(22), 4997–5011 (2007)
[4] Min, F., Liu, Q., Fang, C.: Rough sets approach to symbolic value partition. International Journal of Approximate Reasoning 49, 689–700 (2008)
[5] Zhu, W.: Relationship among basic concepts in covering-based rough sets. Information Sciences 179(14), 2478–2486 (2009)
[6] Min, F., He, H., Qian, Y., Zhu, W.: Test-cost-sensitive attribute reduction. To Appear in Information Sciences (2011)
[7] Min, F., Liu, Q.: A hierarchical model for test-cost-sensitive decision systems. Information Sciences 179(14), 2442–2452 (2009)
[8] Yao, Y.Y.: Granular computing: basic issues and possible solutions. In: The 5th Joint Conference on Information Sciences, vol. 1, pp. 186–189 (2000)
[9] Yao, J.T., Yao, Y.Y.: A granular computing approach to machine learning (2002), http://www2.cs.uregina.ca/jtyao/Pagers/GrcMining-1534.pdf
[10] Yao, J.T., Yao, Y.Y.: Induction of Classification Rules by Granular Computing. In: Alpigini, J.J., Peters, J.F., Skowron, A., Zhong, N. (eds.) RSCTC 2002. LNCS (LNAI), vol. 2475, p. 331. Springer, Heidelberg (2002)

[11] Liu, B., Hsu, W., Ma, Y.: Integrating classification and association rule mining. In: KDD, pp. 80–86 (1998)

[12] Dong, G., Zhang, X., Wong, L., Li, J.: CAEP: Classification by Aggregating Emerging Patterns. In: Arikawa, S., Nakata, I. (eds.) DS 1999. LNCS (LNAI), vol. 1721, pp. 30–42. Springer, Heidelberg (1999)

[13] Li, W., Han, J., Pei, J.: CMAR: accurate and efficient classification based on multiple class-association rules. In: ICDM, pp. 369–376 (2001)

[14] Cover, T., Hart, P.: Nearest neighbor pattern classification. IEEE Transactions on Information Theory 13, 21–27 (1967)

[15] Dasarathy, B.: Nearest Neighbor Pattern Classification Techniques. IEEE Computer Society Press (1990)

[16] Bian, H.Y.: Fuzzy-rough nearest neighbor classification: an integrated framework. In: Proceedings of IASTED International Symposium on Artificial intelligence and Applications, pp. 160–164 (2002)

[17] Guo, G., Wang, H., Bell, D.A., Bi, Y., Greer, K.: An kNN model-based approach and its application in text categorization. In: Gelbukh, A. (ed.) CICLing 2004. LNCS, vol. 2945, pp. 559–570. Springer, Heidelberg (2004)

[18] Neskovic, P., Cooper, L.N.: Improving nearest neighbor rule with a simple adaptive distance measure. Pattern Recognition 28(2), 21–27 (2007)

[19] Yin, X., Han, J.: CPAR: classification based on predictive association rules. In: SDM, pp. 331–335 (2003)

[20] Quinlan, J.R., Cameron-Jones, R.: FOIL: A Midterm Report. In: Brazdil, P.B. (ed.) ECML 1993. LNCS, vol. 667, pp. 3–20. Springer, Heidelberg (1993)

[21] Cohen, W.: Fast effective rule induction. In: ICML, pp. 115–123 (1995)

[22] Cendrowska, J.: PRISM: An algorithm for inducing modular rules (1988)

[23] Clark, Boswell, R.: Rule induction with CN2:some recent improvements. In: EWSL, pp. 151–163 (1991)

Weighting Method Based on Entropy Analysis for Multi-sensor Data Fusion in Wireless Sensor Networks

Donghyok Suh[*], Shinsook Yoon, Seoin Jeon, and Keunho Ryu

Department of Computer Science, Chungbuk National University, Chungbuk, Korea
hanhwaco@kdu.ac.kr, yss28@daum.net,
jsi0198@korea.com, khryu@chungbuk.ac.kr

Abstract. The study of data processing for wireless sensor networks has an interest in filtering, aggregation, and data fusion, and additionally has tended to focus on power reduction in the network. To access the data in a real context at the higher level, the network should consist of heterogeneous multi-sensors, and should converge for the multi-sensors data, which has been sent from the heterogeneous sensors. In this paper, a weighting method based on the sensors has been proposed dependent on the fusion of the multi-sensor data of wireless sensor network. This is based on Dempster-Shafer's evidence theory.

At this point, the valid entropy weighting method has been introduced as a rational weighting method, assuming the circumstances to be identified have been influenced by multiple factors. The data has been fused after weighing on the basic probability assignment function for each sensor, following the weighing method. The contexts have been induced after weighting, and compared to the contexts prior to weighting.

Keywords: Data fusion, weighting, multi-sensor, Dempster-Shafer theory.

1 Introduction

A wireless sensor network is developed to eventually to acquire the real contexts of data, and to provide an intelligent personalization service. The multi-sensor data fusion has increasingly focused on acquiring more-improved contexts[7][8][9].

It is necessary for the sensors of terminal nodes to consist of the different sensors to acquire the real contexts, which are aimed at by wireless sensor network, as better contexts can be induced by using the contexts' data, collected by heterogeneous sensors, rather than by collecting it by a single sensor. For specific examples, it has been shown that the smoke from a fire can be detected, but it cannot be distinguished from fog in daylight, for example in the image detection system for a forest fire. In this case, it could be possible to distinguish the smoke situation and the fog situation if fusing occurs with the contexts collected by a humidity sensor. In the field of data processing for a wireless sensor network, the necessity of the study for the multi-sensor data fusion has been recognized and more focused.

In the existing studies, each sensor has been mostly presumed as having equal importance. The difference between the circumstances of data collected by each

---

[*] Corresponding author.

T.-h. Kim et al. (Eds.): DTA/BSBT 2011, CCIS 258, pp. 41–50, 2011.

sensor and the deduction of the contexts has shown differences in the level of contribution or importance, in reality. Therefore, it is necessary to seek the rational weighting method to deduce the explicit and high-quality contexts. In this paper, the multi-sensor data has been fused, based on the Dempster-Shafer's evidence theory, and the weighting method based on the basic probability assignment function values for each sensor has been proposed.

This paper aims to use the weighting method on the basic probability assignment function for each sensor, and to deduce the high quality contexts by reflecting it on the multi-sensor data fusion. This paper is divided as follows. In Section 2, the theoretical basis has been arranged for the data fusion. In Section 3, the multi-sensor data fusion is conducted, based on Dempster-Shafer theory, and the weighting method based on the Entropy analysis method has been proposed as the rational weighting method for it. In Section 4, an evaluation and conclusion is stated, comparing the weighting methods pertaining to each frequency.

2   Multi-sensor Data Fusion

2.1   Theoretical Background for the Multi-sensor Data Fusion

The Dempster-Shafer Evidence theory calculates the Basic Probability Assignment, when the independent Basic Probability Assignments have been defined[10][11][12]. Provided that Θ is defined as a universal set, which consists of the exclusive propositions. and 2^{Θ} is the power set of Θ, incorporating all the possible combination of the propositions, m could have the four features, is shown below.

$$m : 2^{\Theta} \to [0,1], \; m(\varnothing) = 0, \; 0 \leq m(A) \leq 1 \text{ and } \sum_{A \subseteq \Theta} m(A) = 1$$

If the Basic Probability Assignments have been defined to meet the above four features, the fusion can be possible. The value bel(A), is a belief function equivalent to the assignment values of the quantified total belief for combination A of the individual propositions is defined, as shown below.

$$bel(A) = \sum_{X \subset A} m(X), \text{ for all } A \subseteq \Theta$$

The Basic Probability Assignments function is calculated based on the values of the total fusion system. Fusing the Basic Probability Assignments function values, m1,m2,...,mn, independently calculated from n modules, yields the following.

$$m(A) = \frac{\sum_{B_1 \cap \cdots \cap B_n = A} m_1(B_1) m_2(B_2) \cdots m_n(B_n)}{1-k} = \frac{\sum_{B_1 \cap \cdots \cap B_n = A} m_1(B_1) m_2(B_2) \cdots m_n(B_n)}{\sum_{B_1 \cap \cdots \cap B_n = \varnothing} m_1(B_1) m_2(B_2) \cdots m_n(B_n)}$$

Here, $k = \sum_{B_1 \cap B_2 \cap \cdots \cap B_n = \varnothing} m_1(B_1) m_2(B_2) \cdots m_n(B_n)$

2.2 Related Studies

The study for data fusion with introducing Dempster-Shafer Theory has been continuously refined[1][2][3][4][5]. Also, the study for weighting the subject to be fused upon data fusion using Dempster-Shafer Theory has been earlier proposed[6].

Using the Dempster-Shafer Evidence Theory, the defects of a system can be deduced on the basis of various symptoms occurring in the system, and the high-leveled image information can be acquired thought fusing the different images for specific area in the geographic information field[8][13][14]. Specifically, Rakowsky has stated the detailed calculation procedures of the multi-sensor data fusion to deduce the contexts [15].

Sungwon Park et al. has studied the data fusion method using Dempster-Shafter Theory to acquire high recognition rates in the face recognition field. In their study, the fusion system by means of fusing the four independent face recognition modules has been proposed, and they have acquired better results than the single module which has demonstrated the best performance by fusing two to four modules[6].

Huadong Wu et al. have studied the data fusion method to acquire the enhanced contexts, based on wireless sensor networks, and applied the Dempster-Shafer Evidence Theory to data fusion. In their study, the weighting method for the subject for data fusion has also been proposed [7].

3 Weighting in the Multi-sensor Data Fusion

3.1 Entropy Weighting Method

The definition used in Entropy method is shown as follows.

$$S(p_1, \cdots, p_n) = -k \sum_{i=1}^{n} p_i \ln p_i$$

Here, k=positive constant, meaning $1/(\ln m)$, and

$$p_i = \frac{x_i}{\sum_{i=1}^{m} x_i}$$, evaluates values for factor i.

For m alternatives and n factors, the evaluation values of alternative i for factor j is

$$p_{ij} = \frac{x_{ij}}{\sum_{i=1}^{m} x_{ij}}$$, and Entropy is $E_j = -k \sum_{i=1}^{m} p_{ij} \ln p_{ij}$ $(0 \le E_j \le 1)$.

The variable d_i, is the degree of diversification of the provided information by evaluation at factor j with $d_j = 1 - E_j$ $(1 \le j \le m)$ for all j. If weighting is based on the subjective weighting values of specialists, s_i, the method to put the objective weighting value with the standard of practical importance level between factors is

$$w_j^* = \frac{s_j w_j}{\sum_{i=1}^{n} s_i w_i}$$. Here, it can be applied as $$w_j = \frac{d_j}{\sum_{i=1}^{m} d_j}$$.

3.2   Weighing the Multi-sensor Data Fusion for Deducting the Enhanced Circumstances Information

To seek the weighting method, the wireless sensor network system detecting the forest fire can be considered. The collected data from the heterogeneous sensors deployed to each point has been fused by this system. The defined details of the environment in the wireless sensor network system for detecting the forest fire is shown as follows.

1) Type of Available Sensors : Temperature sensor, Illumination sensor, Humidity sensor
2) Configuration of Event for each sensor:
a. In case of detecting temperature changes, and the variable breadth out of the certain range, this information is reported by the temperature sensor.
b. In case of detecting the circumstances with abnormal darkness in daylight, and the circumstances with abnormal brightness in night, this information is reported by the illumination sensor.
c. In case of the measured humidity in air shown under certain values, this information is reported by the humidity sensor.

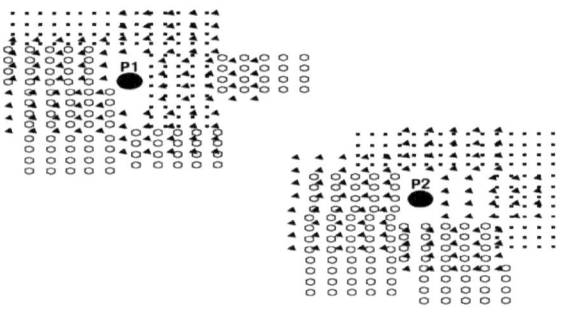

Fig. 1. Scatter Diagram of Sensors for each location

In Figure 1, the temperature sensor, the illumination sensor, and the humidity sensor has been located at P1, P2, and the report from the event of sensors is collected. The Entropy weighting values is applied to the basic probability assignment function for each sensor. The evaluation values for factor I according to each type of

$$p_{ij} = \frac{x_{ij}}{\sum_{i=1}^{m} x_{ij}}$$

sensors, is calculated from the definition of Entropy. Provided that only the 60 temperature sensors out of 300 sensors, including temperature sensors, illumination sensors, and humidity sensors have reported the event situation, Entropy,

$$E_j = -k \sum_{i=1}^{m} p_{ij} \ln p_{ij} \quad (0 \le E_j \le 1)$$

, can be calculated, after applying the factor evaluation values, according to the other type of circumstances as 0.2. Therefore, , as $k=1/(\ln m)$, the Entropy becomes Ej=0.056 with $k=1/\ln 300$. Following this, the degree of diversification of evaluation information provided from each sensor, is calculated

$$w_j = \frac{d_i}{\sum_{i=1}^{m} d_j}$$

as 0.944 with $d_j = 1 - E_j$ $(1 \leq j \leq m)$, and , the weighting values, is calculated as 0.1419 with using each interest factor, d_j.

Table 1. Entropy Weighting Values for each sensor (h1:temperature, h2:Illumination, h3:Humidity)

P1				2^{Ω}	P2			
Number of sensor	Number of event	Contexts of m & factors of n	Entropy weighting		Number of sensor	Number of event	Contexts of m & factors of n	Entropy weighting
100	60	0.2	0.1419	{h1}	100	50	0.1667	0.1433
100	10	0.033	0.1474	{h2}	100	20	0.0667	0.1464
100	30	0.1	0.1443	{h3}	100	30	0.3	0.1416
200	70	0.2333	0.1414	{h1∪h2}	200	70	0.2333	0.1422
200	90	0.3	0.1409	{h1∪h3}	200	80	0.2667	0.1418
200	40	0.1333	0.1433	{h2∪h3}	200	50	0.1667	0.1433
300	100	0.33	0.1407	Ω	300	100	0.3333	0.1415

Belief values (*bel*) and plausibility values (*pl*) according to the basis probability assignment function for each sensor is calculated. The results are shown as Table. 2. The basic probability assignment function at P1 has been indicated as $m(A_k)$, and at P2 indicated as $m(B_k)$.

Table 2. Belief and Plausibility according to the basic probability assignment function for each interest factor(sensor) prior to weighting

$m(A_k)$	$bel(A_k)$	$pl(A_k)$	2^{Ω}	$m(B_k)$	$bel(B_k)$	$pl(B_k)$
0.25	0.25	0.85	{h$_1$}	0.2	0.2	0.8
0.15	0.15	0.75	{h$_2$}	0	0	0.2
0	0	0.15	{h$_3$}	0.2	0.2	0.8
0.45	0.85	1	{h$_1$ ∪ h$_2$}	0	0.2	0.8
0	0.25	0.85	{h$_1$ ∪ h$_3$}	0.4	0.8	1
0	0.15	0.75	{h$_2$ ∪ h$_3$}	0	0.2	0.8
0.15	1	1	∪	0.2	1	1

3.3 Entropy Weighting Values

The weighting value is set as s_i, which is the subjective weighting values of experts according to the Entropy Analysis Method introduced in 3.1. The method to calculates the weighting values with the standard of the degree of the practical importance, which can be shown as $w_j^* = s_j w_j / \sum_{i=1}^{n} s_i w_i$. Based on this, the weighting values calculation method, is calculated by weighing the belief and plausibility for each location, each interest, and each factor after calculating the weighting value for the basic probability assignment function from each interest factor.

Table 3. Reliability and Probability after weighting according to the Entropy Analysis Method for each interest factor (sensor)

$m(A_k)$	$bel(A_k)$	$pl(A_k)$	2^{x^z}	$m(B_k)$	$bel(B_k)$	$pl(B_k)$
0.200	0.200	0.896	{h1}	0.202	0.202	0.801
0.104	0.104	0.800	{h2}	0	0	0.199
0	0	0.099	{h3}	0.199	0.199	0.798
0.597	0.901	1	{h1∪h2}	0	0.202	0.801
0	0.200	0.896	{h1∪h3}	0.400	0.801	1
0	0.104	0.800	{h2∪h3}	0	0.199	0.798
0.099	1	1	Ω	0.199	1	1

On the basis of adjusting the basic probability consignment function values for each interest factor after weighting, the multi-sensor data fusion is processed. After normalizing the basic probability consignment functions of each sensor and finally completing the uncertainty sector, based on reliability and probability of the interest factors through the simplification procedure, the results are shown in the following table.

Table 4. Results of Belief and Uncertainty Sector after Weighting Entropy Weighting Values

	m	bel	Cmn	pl	pl-bel
Ω	0.0260	1	0.0260	1	0
{h1∪h2}	0.1571	0.8957	0.1831	0.9739	0.0783
{h1∪h3}	0.0522	0.7896	0.0783	0.9727	0.1831
{h1}	0.7113	0.7113	0.9466	0.9466	0.2353
{h2}	0.0273	0.0273	0.2104	0.2104	0.1831
{h3}	0.0261	0.0261	0.1043	0.1043	0.0783

We can deduce the circumstances on the basis of the final results table. The interest factors in the table are related to the hypothesis. The hypothesis with the most differential value can be found by comparing the belief values of the hypothesis. In addition, the hypothesis which has the most significant influence on the circumstances, among the hypothesis with bare differential values, can be selected by comparing each uncertainty sector. In the next section, contribution of weighting to the circumstances deduction is evaluated by comparing changes after belief, the uncertainty sector and the weighting of each hypothesis prior to weighting.

4 Comparison and Evaluation

In this section, the differential between the weighting method based on the Entropy Analysis Method and other calculation of weighting value and results of weighting is compared and evaluated.

The weighting values based on the frequency of events are simply calculated based on the weighting values with the standard of the frequency of events, rather than rationally calculating the sophisticated factors, as deducting the circumstances information of the environment, in which sensors have been positioned, or of objects, on which sensors have been mounted. Differing from this, the weighting values based on the Entropy Analysis Method are calculated based on various factors reasonably, which have influence on the circumstances information.

Table 5. Comparison of the Results prior and after weighting

(a) Prior to weighting.

	m	bel	cmn	pl	pl-bel
Ω	0.0405	1	0.0405	1	0
{h1∪h2}	0.1216	0.8378	0.1622	0.9595	0.1216
{h1∪h3}	0.0811	0.7973	0.1216	0.9595	0.1622
{h1}	0.6757	0.6757	0.9189	0.9189	0.2432
{h2}	0.0405	0.0405	0.2027	0.2027	0.1622
{h3}	0.0405	0.0405	0.1622	0.1622	0.1216

(b) After weighting (weighting values based on the frequency of events)

	m	bel	cmn	pl	pl-bel
Ω	0.0181	1	0.0181	1	0
{h1∪h2}	0.1155	0.9220	0.1336	0.9837	0.0617
{h1∪h3}	0.0436	0.8609	0.0617	0.9945	0.1336
{h1}	0.8010	0.8010	0.9782	0.9782	0.1772
{h2}	0.0055	0.0055	0.1391	0.1391	0.1336
{h3}	0.0163	0.0163	0.0780	0.0780	0.0617

(C) After weighting (weighting values based on Entropy Analysis)

	m	bel	cmn	pl	pl-bel
Ω	0.0260	1	0.0260	1	0
{h1∪h2}	0.1571	0.8957	0.1831	0.9739	0.0783
{h1∪h3}	0.0522	0.7896	0.0783	0.9727	0.1831
{h1}	0.7113	0.7113	0.9466	0.9466	0.2353
{h2}	0.0273	0.0273	0.2104	0.2104	0.1831
{h3}	0.0261	0.0261	0.1043	0.1043	0.0783

In Table.5 (a), belief, plausibility, and uncertainty sector prior to weighting are shown. The weighting values based on the frequency of events have been set on the basic probability assignment function(m) of each hypothesis, equivalent to each interest factors, and the belief, plausibility, and uncertainty sector of the interest factors have been calculated based on these values.

These results can be compared to the values prior to weighting shown in Table.5 (a). According to the comparison of results between (a) and (b), the belief and plausibility of the interest factors $\{h_1\}$, $\{h_1 \cup h_2\}$, $\{h_1 \cup h_3\}$ after weighting has increased, and the uncertainty sector has been reduced. The results of weighting based on Entropy Analysis are shown in Table.5 (c). Seen as this results, belief and plausibility of $\{h_1 \cup h_2\}$ among the interest factors have increased and the uncertainty sectors have been reduced, while the interest factor $\{h_1\}$ has been shown to influence on the circumstance deduction as uncertainty sectors have been enlarged. Comparison of the above table with the chart in shown in Figure 2. .

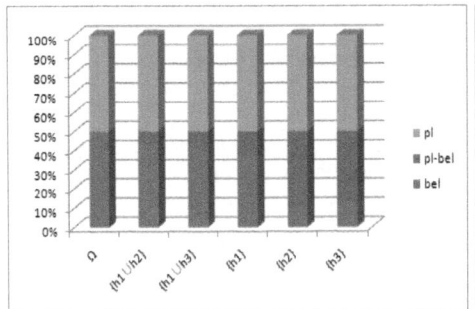

(a) Prior to weighting.

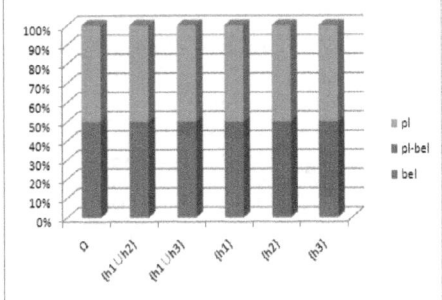
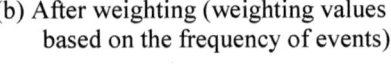

(b) After weighting (weighting values based on the frequency of events)

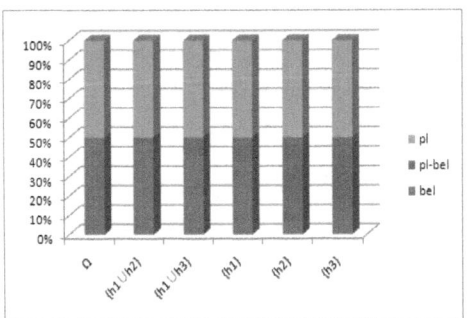

(c)After weighting (weighting values based on Entropy Analysis)

Fig. 2. Comparison of results prior and after weighting

In the deduction of the circumstances, the belief(*bel*) and uncertainty sector(*pl-bel*) is considered as a crucial foundation.

Upon setting the weighting values based on the events, the belief of the interest factors has ironically shown a slight decrease. However, it has contributed to the circumstances deduction as the uncertainty sector has been more enlarged, which has been caused from the circumstance deduction by comparing the uncertainty intervals with the same belief. Compared to this, the result of weighting based on the Entropy Analysis has shown the explicit differentials among the interest factors. In other words, the interest factors with the high belief, the belief of $\{h_1 \cup h_2\}$, $\{h_1 \cup h_3\}$, $\{h_1\}$ have increased.

On the other hand, the belief of the other interest factors has been seen to lower. This result can be helpful to distinguish the hypothesis to be selected explicitly as deducing the circumstances. Therefore, it is helpful to put the weighting values for the circumstances deduction, so that the calculation of the weighting values can be seen to contribute to induce the reasonable results based on the Entropy Analysis Method in the calculation method of the weighting value.

5 Conclusion and Further Study

In this paper, fusing the heterogeneous event information using the Dempster-Shafer Evidence Theory demonstrates the necessity of setting the weighting values on circumstances data according to the sensors subject to fusion, has been scrutinized.

At this point, the fusion has been processed after setting the weighting values on the circumstances data according to sensors subject to fusion by introducing the weighting method, following the Entropy Analysis Method. It can be especially useful for researching the differentials between data as a method which has been frequently used in the case with metric type of data.

Considering that the belief values are regarded as an important factor for deducting the circumstance of evaluation results, it can be confirmed that the values are helpful for deducting the circumstance information by enlarging the differentials among the interest factors. In further studies, it is necessary to apply it to the circumstances recognition in reality, in other various cases.

Acknowledgments. This work was supported by the National Research Foundation of Korea(NRF) grant funded by the Korea government(MEST) (No. 2011-0001044).

References

[1] Malpica, J.A., Alonso, M.C., Sanz, M.A.: Dempster-Shafer Theory on geographic information systems: A survey. Expert Systems with Applications 21(1) (2007)

[2] Kim, P.: A Study on the Performance Improvement of Rocchio Classifier with Term Weighting Methods. Korea Society for Information Management 25(1), 211–233 (2008)

[3] Kim, Y.: An Efficient Parametric Algorithm based Target Classification Scheme (PATaCS) in Wireless Sensor Networks. Korea Information & Communication Graduated School, Docter degree article (2009)

[4] Eduardo, F.N., Antonio, A.F., Alejandro, C.F.: Information fusion for Wireless Sensor Networks: Methods, Models, and Classifications. ACM Computing Surveys 39(3), article 9 (August 2007)

[5] Lohweg, V., Monks, U.: Sensor fusion by two-layer conflict solving. In: 2nd International Workshop on Cognitive Information Processing (CIP), June 14-16, pp. 370–375 (2010)

[6] Park, S., Kwon, J., Choi, J.: 2D Face Image Recognition and Authentication Based on Data Fusion. Korea Institute of Intelligent Systems 11(4), 302–306 (2001)

[7] Wu, H., Siegel, M., Ablay, S.: Sensor Fusion using Dempster-Shafer Theory II: Static Weighting and Kalman Filter-like Dynamic Weighting. In: Proceedings IEEE Annual Instrumentation and Measurement Technology Conference, IMTC 2003, Vail, COUSA, May 20-22 (2003)

[8] Koks D., Challa S.: An Introduction to Bayesian and Dempster-Shafer Data Fusion. DSTO Systems Sciences Laboratory, Commonwealth of Australia (2005)

[9] Kang, M.: Automatic document classification using weight association rule algorithm., Thesis of M.A, Seoul Women's University (2003)

[10] Dempster, A.P.: New Methods for Reasoning towards Posterior Distributions based on Sample Data. The Annals of Mathematical Statistics 37, 355–374 (1966)

[11] Dempster, A.P.: Upper and Lower Probablities Induced by a Multivalued Mapping. The Annals of Mathmatical Statistics 38, 325–339 (1967)

[12] Shafer, G.: A Mathmatical Theory of Evidence. Princeton University Press, Princeton (1976)

[13] Hollnagel, E.: Cognitive Reliability and Error Analysis Method - CREAM. Elsevier, Amsterdam (1998)

[14] Nuclear Regulatory Commission, Technical Basis and Implementation Guidelines for a Technique for Human Event Analysis (ATTEANA), NUREG-1624 (1999)

[15] Rakowsky, U.: Fundamentals of Dempster-Shafer theory and its applications to system safety and reliability modeling. RTA, #3–4 (2007)

A Forecasting Model for Technological Trend Using Unsupervised Learning

Sunghae Jun

Depatment of Bioinformatics and Statistics, Cheongju University,
360764 Chungbuk, Korea
shjun@cju.ac.kr

Abstract. Many results of the developed technologies have applied for patents. Also, an issued patent has exclusive rights granted by a government. So, all companies in the world have competed with one another for their intellectual property rights using patent application. Technology forecasting is one of many approaches for improving the technological competitiveness. In this paper, we propose a forecasting model for technological trend using unsupervised learning. In this paper, we use association rule mining and self organizing map as unsupervised learning methods. To verify our improved performance, we make experiments using patent documents. Especially, we focus on image and video technology as the technology field.

Keywords: Image and video technology, Technology forecasting, Unsupervised learning, Association rule mining, Self organizing map.

1 Introduction

In this paper, we propose a forecasting model for technological trend using unsupervised learning. Association rule mining (ARM) and self organizing map (SOM) are popular methods of unsupervised learning. These are used for constructing our forecasting model. Also, we focus on image and video technology (IVT) as a given technological field for our case study. IVT of multimedia database systems has been used an important tool for providing information to humans [1-2]. Many researches of coding, processing, visualization, analysis, and retrieval have been developed in IVT. Recently, biometrics and forensics are needed the IVT [3]. In general, the results of researched and developed technologies for IVT have been published as patent and paper. These are massive literatures. It is difficult to construct forecasting model by them using the quantitative methods of statistics and machine learning [4]. Most technology forecasting (TF) models have been depended on the qualitative methods such as Delphi [5-8]. These qualitative TF are not stable because they are based on subjective knowledge of domain experts. So, we need more objective TF method for IVT forecasting. In this paper, we use a combine model by ARM and SOM to construct a quantitative method for IVT forecasting. ARM is a popular predictive method based on conditional probability [9-10]. SOM is a typical clustering model in unsupervised neural networks of machine learning [11-12]. ARM

T.-h. Kim et al. (Eds.): DTA/BSBT 2011, CCIS 258, pp. 51–60, 2011.

and SOM will create a synergy effect each other to forecast IVT effectively. We will use patent document about IVT until now for IVT forecasting. By analyzing the IVT patent documents, we forecast especially vacant areas of IVT using patent data of U.S, Europe, and China from USPTO (United State Patent and Trademark Office, www.uspto.gov). In next section, we review the related works of IVT and TF. We introduce proposed method to forecast vacant technology of IVT in section 3. To verify the performance our work, we will show experimental results in section 4. Final section includes our conclusions and future works.

2   Image and Video Technology Forecasting

Image technology includes the techniques of material application and management method to create, storage, and analyze images. Based on the image technology, video technology is to capture, record, manage, analyze and reconstruct the continuous images of motion [1-3]. So, IVT is a combined technology based on image and video techniques. TF is to predict a moving trend of technological change. We have R&D plan to avoid infringement of intellectual property using TF results. TF also support meaning knowledge for technology marketing and reducing risk of R&D investment in company and government [13-15]. In this paper, we find vacant TF to give company and government the technological feasibility of each vacant aspect in IVT. The important goal of TF is to monitor the technological trend of given technology [15]. Most TF works have used qualitative and subjective methods depended on experts' prior knowledge such as Delphi [5-8]. TF results of the method are not stable. So, we need more quantitative and objective methods than previous approaches [14]. Some approaches such as roadmap and bibliometrics, have been introduced for TF [14],[16-17]. Typical data source for constructing TF model are paper and patent databases. We use applied patent documents of IVT to forecast vacant technology of IVT in this paper. Principal component analysis (PCA) and SOM were the methods studied on previous TF researches [14],[18-19]. In this paper, we propose a combined TF model including ARM and previous TF method by SOM. We will improve the performance of vacant TF using the proposed methods.

3   IVT Forecasting Using ARM and SOM

Retrieved patent data are so large. We need an analytical method for large patent documents. ARM is a popular method for discovering knowledge from large databases. ARM mines frequent item sets to find meaningful relationship between variables. ARM analyzes item and transaction data sets [9]. $I=\{i_1, i_2, ..., i_n\}$ is a set of n binary items. $T=\{t_1, t_2, ..., t_m\}$ is a set of transactions. A transaction of T consists of unique identical number and items. A rule of ARM is represented as follows.

$$X \rightarrow Y \quad (X \cap Y = \phi) \tag{1}$$

Where, X and Y are in I. Also X and Y are antecedent and consequent of the rule. We can get so many rules from ARM results. To select meaningful rules from all possible rules, we need criteria to evaluate all rules. Support and confidence are well known constraints. We select the rules satisfied minimum threshold on support and confidence. Support is defined as follow.

$$\text{support}\,(X \rightarrow Y) = \frac{nmber\ of\ transactions\ containing\ both\ X\ and\ Y}{total\ number\ of\ transactions} \tag{2}$$

So, support$(X \rightarrow Y)$ is equal to probability $P(X \rightarrow Y)$. confidence$(X \rightarrow Y)$ is represented by probability $P(Y/X)$ as follow.

$$\text{confidence}(X \rightarrow Y) = P(Y \mid X) = \frac{P(X \cap Y)}{P(X)} = \frac{\text{support}(X \rightarrow Y)}{\text{support}(X)} \tag{3}$$

Lift is additional interest measure in ARM. We can filter the rules satisfied minimum support and confidence constraints to select more meaningful rules [14]. This is defined as follow.

$$\text{lift}(X \rightarrow Y) = \frac{\text{support}(X \Rightarrow Y)}{\text{support}(X)\text{support}(Y)} \tag{4}$$

We get more significant rules by lift measure. Next, the relationship between items is shown by lift value.

$$\text{lift}(X \rightarrow Y)\,value = \begin{cases} > 1, & X\ and\ Y\ are\ complementary \\ 1, & X\ and\ Y\ are\ independent \\ < 1, & X\ and\ Y\ are\ substitutive \end{cases} \tag{5}$$

When the lift value equals to 1, item X and Y are independent each other. That is, X and Y are not associated. The relationship between X and Y is more complementary according as the value is increased over 1. Also, X and Y are substitutive if the value is under 1. In this paper, we use support, confidence, and lift as measures for finding meaningful rules to forecast the IVT. Retrieved patent documents and extracted terms are used as transactions and items of ARM. From the ARM results, we get the relationships between detailed technologies of IVT. Self organizing map (SOM) is a competitive neural networks for classification and clustering [19]. SOM consists of two layers, input and feature. We cluster all patent documents to 2×2 feature map. First, SOM normalize input vector x and initialize weight matrix M. Second, we compute the distance between input and weight using Euclidean distance. The closest m_j to x_i is updated as follow [20].

$$m_k = m_j + \alpha(x_i - m_j) \qquad (6)$$

Where, m_j and m_k are current and new weights. So m_k moves to x_i. The constant α is a learning rate to control the speed of converging optimal point. SOM clustering is repeated until satisfying given conditions. We cluster all patent documents to feature map. We find the vacant technology in the feature map with assigned patent documents. In the feature map of SOM, the areas with low density are defined as vacant technology.

In this paper, we use ARM and SOM for forecasting technology of IVT. Our proposed TF approach consists of four steps. First, we prepare data set for constructing IVT forecasting model. We prepare IVT patent documents as analyzed data by ARM and SOM. We retrieve IVT patent documents from USPTO. There is a problem in first step. In general, patent documents are not suitable to analytical methods of statistics and machine learning such as ARM and SOM. To solve this problem, we preprocess the documents using text mining techniques. The output of first step is a document-term matrix (DTM) preprocessed. Our DTM is a structured data to be analyzed. Second, ARM is used for IVT forecasting. We extracted top ranked keywords from DTM, output of first step. Patent document and keyword are transaction and item for ARM. We construct a set of ARM rules by given support and confidence values. In this paper, we determine the values as small as possible because we wish to get ARM rules as many as possible. We search meaningful rules from the set of ARM rules and determine three rules with the highest lift, confidence, and support values. These rules are used to forecast IVT. Third, we use SOM to cluster IVT patent documents. The documents are assigned to the feature map of SOM. In general, the number of clusters is proportional to the dimension of feature map. We perform SOM clustering from large dimension of the feature map to small dimension. In the last SOM result, we find a cluster with vacant technology. The cluster is relatively small but not sparse. Top ten terms form the patent document of the cluster define as vacant area are extracted for defining the vacant technology for IVT. Fourth, we forecast the technology for IVT using the outputs of second and third steps. The following process shows our proposed model for IVT forecasting step-by-step.

Technology Forecasting for Image and Video Technology
  Step1. Preparing data set for forecasting IVT
    (1-1) Determining keywords equation for retrieving;
    (1-2) Retrieving IVT patent documents from USPTO;
    (1-3) Preprocessing documents using text mining;
    (1-4) Getting document-term matrix from (1-3)
result;
    (1-5) Dividing DTM into training and test data sets;
    (Output) DTM divided into training and test data.
  Step2. Extracting ARM rules
    (2-1) Extracted top ranked keywords from output of
        step1;
    (2-2) Constructing ARM rules;
    (2-3) Searching meaningful rules by ARM criteria;
      Rule1 by the heighted lift value,
      Rule2 by the heighted confidence value,

```
           Rule3 by the heighted support value.
    (Output) Three ARM rules - Rule1,2,3.
Step3. Clustering DTM
    (3-1) Performing SOM clustering from large dimension
          of feature map to small dimension of feature
          map;
    (3-2) Determining optimal number of clusters;
    (3-3) Finding a cluster with vacant technology;
    (3-4) Extracting top ten terms from the patent
          documents included to vacant area;
    (Output) Ten terms extracted from patent documents
in
          the cluster as vacant technology area.
Step4. Forecasting technology related to IVT using
       outputs from step2 and step3;
```

Also, next figure simplifies the proposed process. A rectangle is an output of each step and an arrow represents the analytical work between outputs.

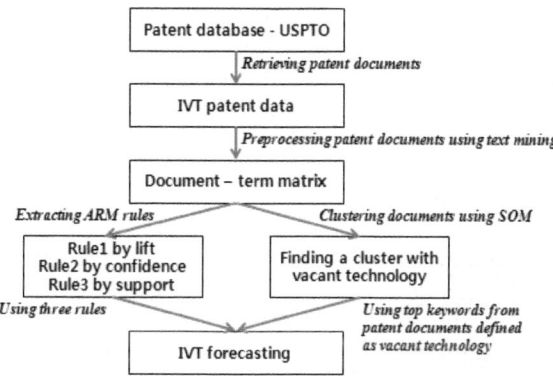

Fig. 1. Proposed IVT forecasting model

To forecast IVT, we combine ARM and SOM results. ARM gives the relationship between technological terms of IVT. Vacant technology of IVT is defined by SOM result.

4 Experimental Results

To forecast vacant technology of IVT, we retrieved patent data from USPTO using the following keyword equation.

*Title = [(image + video) * (technology + coding + processing + analysis + retrieval + forensic + application)]*

Where '+' and '*' are 'or' and 'and' operators, respectively. At the first, we retrieved 5001 patent documents until June 8, 2011. The data included some patent documents not related to IVT. We got 3780 patent documents after removing the patents not related to IVT. The available percentage of our retrieved patent documents was

75.58%. The first applied patent was shown in 1984. The issued IVT technologies as patent documents were increased in the late 1990s. The rate of increase has been fast recently. But, 2007 through 2010, the number of patents of IVT was decreased. We intend to find this trend of IVT from our experiment. We divided all 3780 patent documents into two data sets as follow.

Table 1. Dividing all available patent documents into training and test data sets

Data set	Year	Number of patents
Training	1984 – 2005	2251
Test	2006 – 2010	1529
Total	1984 – 2010	3780

Table 2. Occurred frequency levels of top 30 terms

Terms	Occurred frequency levels		
	Low (0 – 1)	Middle (2 – 3)	high (4 –)
allocation	2249	1	1
audio	2227	5	19
beam	2247	2	2
binary	2215	23	13
calibration	2245	4	2
chromaticity	2237	2	12
conversion	2141	66	44
diffusion	2237	11	3
digital	2174	50	27
document	2188	27	36
electrifying	2250	0	1
gradation	2202	24	25
image-capturing	2242	3	6
lens	2240	10	1
macro	2243	3	5
motion	2203	25	23
multivalue	2250	0	1
object	2116	90	45
quantization	2221	25	5
radiographic	2243	4	4
retrieval	2219	13	19
roller	2249	0	2
scaling	2245	4	2
signal	1948	105	198
synchronization	2246	2	3
texture	2241	8	2
transfer	2227	18	6
transparency	2244	0	7
ultrasonic	2249	0	2
ultrasound	2247	1	3

We constructed a TF model based on ARM using the training data set. To verify the performance of the constructed model, we used test data set. In general, patent document is not suitable for most data analysis methods such as ARM. Text mining is a good preprocessing tool for transforming patent document into structured data[4]. We use 'tm' R package for text mining[20]. This package provides a text mining framework based on R. R is a language for statistical computing[21]. First, we got DTM using text mining preprocessing. The dimension of DTM was 2251×8512. That is, the numbers of documents and terms were 2251 and 8512 respectively. A element e_{ij} of DTM represents the occurred frequency of term j in document i. There were many meaningless terms in the 8521 terms. They were 'and', 'the', 'for', and so on. We eliminated these terms. We also removed common terms such as 'image', 'video', and 'technology'. Second, we extracted top 30 terms from DTM. Next table shows the terms and their occurred frequency levels.

To apply ARM, we replaced occurred term frequency of DTM by occurred level. For example, if the occurred term frequency was 2 or 3, we determined the level was 'middle'. We knew most frequency levels were 'low'. This was a problem for constructing ARM rules. So, we selected the terms with the number of 'row' levels was respectively small. In above table, the words in bold type were the selected terms. We removed the documents (rows) which have all same levels in ten selected terms. For example, 2250th patent document had binary=low, conversion=low, ..., transfer=low. So, we eliminated the document because it was not useful for constructing ARM rules. Third, we found ARM rules in DTM with ten selected terms. The support and confidence were determined as 0.0001 and 0.01 respectively. In this paper, we use 'arules' and 'arulesViz' package of R for our ARM model[22-23]. Finally we got 134,358 ARM rules with support=0.0001 and confidence=0.01. We determined very small values of support and confidence for extracting ARM rules as many as possible. Next, we show four statistics of set of 134,358 rules.

Table 3. Statistics of 134,358 rules

Statistics	Support	Confidence	Lift
Min	0.0009	0.0100	0.1250
Median	0.0045	1.0000	1.0460
Mean	0.0381	0.7554	1.7710
Max	0.9785	1.0000	222.80000

We knew that the average values of support and confidence were large. The average lift value was small relatively. But the maximum value of lift was extremely large. We generated meaningful rules from the set of 134,358 rules. These rules are shown as follow.

We showed top one rule according to each measure for generating ARM rules. First row shows the generated rule with the highest lift value. Second and third rows show the generated rules with the highest confidence and support values. First, in the rule with the highest lift, {(gradation=middle, object=middle)→(quantization=high)}, we found the middle levels of 'gradation' and 'object' were strongly associated with the high level of 'quantization'. Though the occurred frequency of this rule was small, the

Table 4. Generated meaningful rules

{X→Y}	Support	Confidence	Lift
{(gradation=middle,object=middle) → (quantization=high)}	0.0009	1.0000	**<u>222.8000</u>**
{(quantization=high) → → (signal=low)}	0.0049	**<u>1.0000</u>**	1.3736
{(quantization=low) → (transfer=low)}	**<u>0.9515</u>**	1.0000	0.9994

relationship between them was highly correlated. Also, they were complementary each other strongly. We can use this rule to forecast IVT. Second, the occurred probability of low level of ''signal' given the high level of 'quantization' was 1. That is, the low level of 'signal' was depended too much on the high level of 'quantization' from the rule, {(quantization=high)→(signal=low)}. Third, the low levels of 'quantization' and 'transfer' occurred together. In third rule, {(quantization=low)→(transfer=low)}, we found the low level of 'quantization' and the low level of 'transfer' were needed together for developing IVT. We used SOM as another method for IVT forecasting. For our experiment, 'som' package was used[24]. Next figure shows the clustering result of IVT patent documents. Each cell represents the number of assigned patent documents of a cluster.

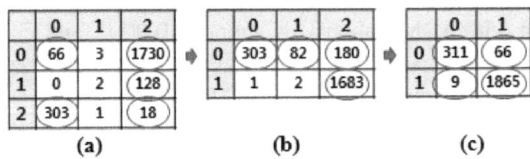

Fig. 2. SOM results of IVT patent documents clustering – (a) 3×3, (b) 3×2, (c) 2×2

First, we determined the dimension of feature map was 3×3 in (a). Maximum number of clusters was nine. We found the proper number of clusters was five, red circles in (a). Second, we did SOM clustering with 3×2 feature map as (b). In this result, we found the number of clusters was four. Third, we clustered IVT patent documents by 2×2 SOM in (c). We concluded the optimal number of clusters was four. In result (c), we selected a vacant area for IVT forecasting in four clusters. (row, column)=(1,1) and (0,0) were not vacant areas because they had so many patent documents. The cluster, (row, column)=(1,0) had only nine patent documents. We decided the technology of this cluster was an outlier field of IVT. Finally, we also decided (row, column)=(0,1) cluster as a vacant technology area for IVT. This cluster had sixty-six patent documents. This was 2.93% of all training data. To define detailed technology represented by (0,1) cluster, we extracted top ten keywords from the sixty-six patent documents. They were 'distortion', 'luminance', 'signal', 'storage', 'intensity', 'reproduction', 'compress', 'digital', 'quantization', and 'transfer'. We can define the vacant technology of IVT using these ten terms.

Next, to verify the performance of our forecasting method, we used test patent documents. The test data had 1529 patent documents from 2006 to 2010. We found 255 patent documents in the test patent documents. These were 16.68% of all test data. They included ten keywords determined by vacant technology for IVT in the training data. So, we knew that the vacant technology for IVT was increased from training data (1984-2005) to test data (2006-1010). It was because a vacant technology has a chance of increasing in future.

5 Conclusions and Future Works

In this paper, we proposed a forecasting method for IVT vacant technology. We used ARM and SOM for our forecasting model. Patent documents related to IVT were retrieved from USPTO. The patent data had all patents of IVT in U.S., Europe, and China. We forecasted the vacant areas of IVT by constructing forecast models using the patent data. In the ARM, we extracted top keywords from IVT patent documents. These terms were used to find the ARM rules for vacant IVT forecasting. According to the levels of the terms, 'gradation', 'quantization', 'object', 'signal', and 'transfer', we found three ARM rules for IVT forecasting. We used SOM for another vacant TF model of IVT. From large dimension of feature map to small feature map, we searched optimal clustering result of IVT patent documents. We concluded four clusters as optimal clustering. A vacant technology area was decided from the four clusters. We determined a cluster with sixty-six patent documents. This cluster was relatively small but not sparse. We extracted top ten terms from the patent documents belong to the cluster defined as a vacant technology. The percentage of the patents defined vacant area was 2.93% in training patent documents, 1984 to 2005. We knew the percentage of our defined vacant technology in test patents, 2006 to 2010 was 16.68%. So, we verified the forecasting performance of our model.

In this paper, we used patent documents for constructing IVT forecasting models. The other important source containing researched and developed results of IVT was in papers published in journal or conference. One of our future works is to make more accurate TF model of IVT using patents and papers together. A limitation of our work was a shortage of IVT domain experts' support. We needed their knowledge to deploy our experimental results to IVT forecasting.

References

1. Amin, T., Zeytinoglu, M., Guan, L.: Application of Laplacian Mixture Model to Image and Video Retrieval. IEEE Transaction on Multimedia 9(7), 1416–1429 (2007)
2. Okamoto, H., Yasugi, Y., Babaguchi, N., Kitahasui, T.: Video Clustering using Spatio-Temporal Image with Fixed Length. In: IEEE International Conference on Multimedia and Expo., pp. 53–56 (2002)
3. Han, J., Kamber, M.: Data Mining Concepts and Techniques. Morgan Kaufmann (2001)
4. Tseng, Y.H., Lin, C.J., Lin, Y.I.: Text mining techniques for patent analysis. Information Processing & Management 43, 1216–1247 (2007)
5. Madu, C.N., Kuei, C.H., Madu, A.N.: Setting priorities for IT industry in Taiwan-A Delphi study. Long Range Planning 24(5), 105–118 (1991)

6. Mitchell, V.W.: Using Delphi to Forecast in New Technology Industries. Marketing Intelligence & Planning 10(2), 4–9 (1992)
7. Woundenberg, F.: An evaluation of Delphi. Technological Forecasting and Social Change 40, 131–150 (1991)
8. Yun, Y.C., Jeong, G.H., Kim, S.H.: A Delphi technology forecasting approach using a semi-Markov concept. Technological Forecasting and Social Change 40, 273–287 (1991)
9. Agrawal, R., Imielinski, T., Swami, A.: Mining Association Rules between Sets of Items in Large Databases. In: Proceedings of the 1993 ACM SIGMOD International Conference on Management of Data, pp. 207–216 (1993)
10. Hahsler, M., Grun, B., Hornik, K.: arules – A Computational Environment for Mining Association Rules and Frequent Item Sets. Journal of Statistical Software 14(15), 1–25 (2005)
11. Kohonen, T.: Self-Organizing Maps. Springer (2000)
12. Hastie, T., Tibshirani, R., Friedman, J.: The Elements of Statistical Learning – Data Mining, Inference, and Prediction. Springer (2001)
13. Metaxiotis, K., Psarras, J.: Expert systems in business: applications and future directions for the operations researcher. Industrial Management & Data Systems 103(5), 361–368 (2003)
14. Yoon, B., Park, Y.: Development of New Technology Forecasting Algorithm: Hybrid Approach for Morphology Analysis and Conjoint Analysis of Patent Information. IEEE Transactions on Engineering Management 54(3), 588–599 (2007)
15. Zhu, D., Porter, A.L.: Automated extraction and visualization of information for technological intelligence and forecasting. Technological Forecastingand Social Change 69, 495–506 (2002)
16. Coates, V., Farooque, M., Klavans, R., Lapid, K., Linstone, H.A., Pistorius, C., Porter, A.L.: On the future of technological forecasting. Technological Forecasting and Social Change 67, 1–17 (2001)
17. Mann, D.L.: Better technology forecasting using systemic innovation methods. Technological Forecasting and Social Change 70, 779–795 (2003)
18. Jun, S., Park, S., Jang, D.: Forecasting Vacant Technology of Patent Analysis System using Self Organizing Map and Matrix Analysis. Journal of the Korea Contents Association 10(2), 462–480 (2010)
19. Jun, S., Uhm, D.: Patent and Statistics, What's the connection? Communications of the Korea Statistical Society 17(2), 205–222 (2010)
20. Feinerer, I., Hornik, K., Meyer, D.: Text Mining Infrastructure in R. Journal of Statistical Software 25(5), 1–54 (2008)
21. R Development Core Team.: R, A language and environment for statistical computing. R Foundation for Statistical Computing (2011), http://www.R-project.org
22. Hahsler, M., Buchta, C., Gruen, B., Hornik, K.: Package 'arules'. R-project CRAN (2011)
23. Hahsler, M., Chelluboina, S.: Package 'arulesViz'. R-project CRAN (2011)
24. Yun, J.: Package 'som'. R-project CRAN (2010)

Mathematical Model and Analysis of Reliability and Safety of Large Database Systems

Armen G. Bagdasaryan[1] and Tai-hoon Kim[2]

[1] Russian Academy of Sciences, Trapeznikov Institute for Control Sciences,
Profsoyuznaya 65, 117997 Moscow, Russia
abagdasari@hotmail.com
[2] Department of Multimedia, Hannam University,
133 Ojeong-dong, Daedeok-gu, Daejeon, Korea
taihoonn@hnu.ac.kr

Abstract. In this paper we consider and analyze the problems of reliability and safety of database systems operation. One of the methods of improving the reliability and quality of functioning of database systems is the registration of information about events and processes occuring in the system. In this direction, we introduce some new concepts related to the problem, and then develop and analyze a corresponding graph-theoretical mathematical model. In conclusion, several future directions of research are also briefly outlined.

Keywords: database systems, database reliability, safety, graph model, information registration, registration graph, combinatorial optimization.

1 Introduction

The majority of modern operating systems supports the system log that records every job performed in the system. Operating system detects the job beginning and the job identifier, the user name, initiating the task, and also provides the storage of this information in the system log during the time of job processing. Such a record contains various special information about the job, the user, devices involved, current state of the processing procedure, restrictions on system memory, access policy, etc. [1]–[3]. At the same time, the system log does not show the changes in data and the values of the data before and after the change. This does not allow one to use system log to monitor control actions and to analyze different types of deviations from normal functioning. Nowadays, the database management systems have special logging programs [3,4], which provide each operation in the database to be recorded in system log. Generally, all the information on all transactions and operations in the database is recorded in the system log, including operations generated by the system, such as, index updating. This log is typically used to restore the database or some of its parts to the original state after a system failure or after a loss of data in the database, or for any other reason [4]. However, large amount of information collected in the log, lack of systematization when recording information, as well as the lack

T.-h. Kim et al. (Eds.): DTA/BSBT 2011, CCIS 258, pp. 61–70, 2011.

of formal means to ensure effective analysis of this information, is a significant impediment for the use of log to measure the efficiency of the system. Furthermore, modern computing systems do not possess sufficient means to evaluate the possible effects of observed deviations from the system normality, and also they do not have the means of efficient *a posteriori* analysis of various situations in the system.

In this regard, the design of data processing systems and of database reliability models has to be followed by the development of a system of information registration and of the tools of its analysis, which is the subject of this paper.

2   Registration Systems: Requirements and Objectives

Registration of information is an effective method of improving the reliability and quality of functioning of data processing systems. The registration system identifies and records in a special logbook events and processes occurring in the system. This information can then be used to analyze and monitor the actions of users of data processing systems, to identify sources of error in the initial, intermediate and output data, to identify attempts to violate the user regulations in the system, including the breaking of protection mechanisms [5,6], as well as to analyze the efficiency of control decisions [4]–[9]. In fact, without logbook it is impossible to trace the events in database systems in the past and thus to identify and eliminate sources of deviation from the normal operation of the system.

The registration system is a complex of technical, information, software and possibly hardware tools aimed at identifying the status of data processing system at arbitrary time instants.

The main purposes of registration system are:

- detecting the violations of the protection of data processing system or the threat of protection breaking;
- analyzing the performance of data processing system, as well as of the entire database;
- providing the operability of data processing system under failures and faults in the system;
- providing the physical safety of data and software, and identifying the causes of their destruction;
- evaluating the reliability of information, controlling data processing; identification, localization and correction of errors, as well as the analysis of the ways of error spreading and interaction in the system;
- debugging the software and configuring the data processing system;
- monitoring the effectiveness of control and management decisions;
- providing a psychological effect on users as a sanction to increase their responsibility upon operating with the system.
- assisting in the training of working with the system.

Registration systems that implement the full range of the above functions are universal. However, this kind of systems have several drawbacks, such as large amount of resources consumed and the high cost of development and maintenance; long-term development; accumulation of large amounts of data that impedes efficient search and analysis of information on implementing the above functions, the complexity of practial realization. These drawbacks necessitate the development of more effective specialized systems.

We can distinguish two main types of registration systems: deterministic and random.

Definition 1. *The registration system is called* deterministic *if actions and/or events are recorded in certain pre-selected control points using a specific algorithm.*

Definition 2. *The* control point *is a point in data processing system (procedure, operation, information array, data element, etc.), at which the registration of information is performed.*

The recording of information is performed either continuously or at certain predefined time intervals.

One of the most effective methods of registration is the random registration.

Definition 3. *The registration system is called* random *if it means a random control of information element, or control of actions and/or events, occurring in the system, at random intervals.*

The deterministic and random registration systems have the same purposes. One of the advantages of random registration systems is the following. The algorithms of deterministic registration system can be fairly easily determined and used to bypass the security system, whereas in a random registration, even if the algorithm is known, determining the amounts, times and periods of registration is very difficult, since it is governed by random laws. The system of random registration along with the identification of attempts of unauthorized access becomes a strong and powerful factor in preventing violations due to the difficulty in predictability of actions of controlling system.

The main functions of registration system are the following: identifying the important processes occurring in the system that affect the efficiency of system's functioning; maintaining registration archive; delivering the necessary information for solving control problems in accordance with the objectives.

Once the efficiency of database systems decreased, the registration system enables a retrospective analysis [10] and then restoring the system state at the time of fulfillment of such events. The functioning of registration system is also necessary upon analysis of various kinds of conflict situations, hacker attacks on the system, the detection of deliberate attempts of unauthorized access to information.

A specialized archive serves as the information database for registration system. This archive records, with the help of special programs, information about the events, significant from the point of view of system objectives.

The specialized archive provides the implementation of the following tasks:

- collecting, structuring and deployment of information gathered by registration system;
- reliable and long-term storage of information;
- protection of recorded information from unauthorized use;
- providing system administration with access to recorded information;
- information updating.

Information about the events occurring in data processing system is collected by the registration system and recorded in the registration archive for storing and, in case of necessity, for providing access to stored information.

In order to provide reliability and safety of database and data processing system operation, system administrator should have all the information about the events and processes in the system, about the user actions performed, and about the results of these actions [11]–[13]. This information can be obtained by using the registration archive. In this case, registration system accumulates the data about the procedures and information arrays used, and sequences of addressing them, which allows one to identify and analyze the character and type of errors occurred, how these errors spread, and programs and users who attempt to get unauthorized access. Functioning of any registration system involves the use of some additional resources, needed to collect, store, and analyze information obtained in the process of database operation. However, carrying out continuous registration, that is, recording absolutely all processes and events occurring in the database, is usually too costly. Therefore, the design of registration system requires the most efficient use of allocated resources, which act as constraints.

In the next section we are concerned with the problem formulation and with developing the mathematical model of deterministic registration system, capable of implementing the above functions.

3   The Problem Formulation and Mathematical Model

Let us consider a formal description and problem formulation for synthesis of registration system satisfying the requirements outlined in previous section. The data processing system, together with database subsystems, includes software, informational, and technical support, and system users.

We define the structure of data processing system in the form of a processing technology graph $G(X, E)$, where $X = X' \cup X''$ is a set of vertices; $X' = \{x'_1\}$ is a subset of informational elements, $X'' = \{x''_0\}$ is a subset of processing procedures, $X' \cap X'' = \emptyset$; E is a set of directed edges that characterize interrelations between the vertices. Thus, the processing technology graph is a bipartite graph with two types of vertices.

Definition 4. *The vertex of the processing technology graph, which collects the information about events and processes taking place in data processing system, is called the* registration point.

The registration points and connections between them determine the graph $R(A, B)$, which is the subgraph of the graph $G(X, E)$.

Definition 5. *The graph $R(A, B)$ is referred to as the* registration graph.

The synthesis of registration system consists in selection of the most rational registration points, in the sense of the goal set accomplishment. Therefore, this problem is equivalent to that of deriving the graph $R(A, B)$. Each vertex of $R(A, B)$ is assigned a characteristic vector $\overline{C}$ that reflects resource, time, and structural characteristics of the registration method, applied to this vertex; the registration method is regarded as a way of collecting, processing, transmitting, and allocating of information of the required type and volume in the registration archive at given time intervals. Hence, the registration system is characterized by topology of the registration points, and also by the set of registration methods, associated with the given topology.

With the above definitions, the general problem of synthesis of deterministic registration system is formulated as follows.

It is necessary to perform a transformation $G(X, E) \xrightarrow{\pi} R(A, B)$ such that it would provide an extremum of some function ϕ, that is extr $\phi\,(R(A, B))$, where π is the operator of transformation of the processing technology graph $G(X, E)$ into the registration graph $R(A, B)$.

The transformation π can be represented as a consecutive transformation of the graph $G(X, E)$

$$G(X, E) \longrightarrow G^*(X, E) \longrightarrow \ldots \longrightarrow R(A, B),$$

where $G^*(X, E)$ is the acyclic graph without loops corresponding to the graph $G(X, E)$. The graph $G^*(X, E)$ is referred to as the *procedure interrelation graph*.

The transformation of the technology graph $G^*(X, E)$ is based on the use of matrix transformations. Hence, further in the text, π should be understood as the transformation $\pi\colon G^*(X, E) \longrightarrow R(A, B)$.

For simplicity of notations, we will omit the upper index *. Then we have $\pi\colon G(X, E) \longrightarrow R(A, B)$, where $G(X, E)$ is the procedure interrelation graph and $R(A, B)$ is the registration graph.

The important peculiarity that characterize the optimality of selecting π is that the registration graph $R(A, B) = \pi(G(X, E))$ should provide the possibility of unique identification of any path in the graph $G(X, E)$, from input procedures to output ones. In this case, the information accumulated by the registration system will allow the database administrator to analyze both the actions of users in the system and possible ways of propagation of errors over information elements and arrays in the process of functioning of data processing system. The main restrictions imposed when selecting π are restrictions on the allocated resources, on the admissible structure of registration system, and on the volume of information to be registered. This problem is analyzed and solved in the next section.

4 Analysis of the Model and Problem Solution

We assume that resources on registration (expenses on data collection, storage, and analysis) in the information registration system do not have rigid limitations, that is, the restriction is such that it allows us to construct the information registration system, modeled by the registration graph $R(A, B)$, that provides unique identification of all possible paths in the graph $G(X, E)$. In other words, we have to uniquely determine the registration graph $R(A, B)$ from $G(X, E)$.

Suppose that the cost for registration in any vertex of the graph G is a fixed value. This takes place in case of constancy of the volume of information to be registered in every vertex. Then, it would be natural to consider the expenses on implementation of the synthesized registration system as its efficiency criterion. Therefore, the problem is reduced to solving the problem of resource allocation over the vertices of the graph $G(X, E)$ with homogeneous arcs.

Let the vertices of the graph G be labeled such that each vertex is assigned a unique number. Then, $x_i \in X$ denotes the vertex of the graph G, which has the number x_i, and X denotes the set of all these numbers. Further, let the vertices of the graph $G(X, E)$ are weighted in accordance with the function $\mu \colon X \longrightarrow \mathbb{R}_+^1, \mu(x_i) = \mu_i, \forall i = \overline{1, |X|}$, where $|\cdot|$ denotes the cardinality of a set. The weight of each vertex corresponds to the value of expense on the registration of procedure x_i.

Let us divide the set X into two subsets X_1 and X_2 such that $X_1 \cap X_2 = \emptyset$, and $\forall x_i \in X_1 \Rightarrow \deg^-(x_i) = 0$ and $\forall x_i \in X_2 \Rightarrow \deg^+(x_i) = 0$, where $\deg^-, \deg^+$ are the outdegree and indegree of the vertex x_i, respectively; X_1 is the set of sources, X_2 is the set of sinks. Let W is the set of paths from X_1 to X_2 (the set W can be obtained by means of known algorithms having polynomial complexity). Then the path $w \in W$ is uniquely represented by finite sequence of numbers of path's vertices: $w = (x_{i1}, x_{i2}, \ldots, x_{i|w|})$.

Definition 6. *A finite sequence $y = y_{j1}, y_{j2}, \ldots, y_{jm}, \forall y_{jk} \in X$, corresponds to some path $w \in W$, $y \approx w$, if and only if all y_{jk} are found in the finite sequence w and in the same order.*

Let $M \subseteq X$ is a subset of the set of vertices of the graph G. Consider the set of finite sequences Y_M consisting of the elements of the set M.

Definition 7. *A finite sequence $y \subseteq Y_M$ is said to be* complete *if there exists a path $w \in W$ such that $y \approx w$ and $M_y = X_w \cap m$, where $M_y \subseteq M$ is the subset of elements composing y, and $X_w \subseteq X$ is the subset of elements composing w.*

Further in the text, we shall consider only complete finite sequences, which we denote by Y_M.

Let us consider the following class of mappings.

$$y \in y_M \colon W \longrightarrow Y_M; \quad \forall w \in W \; \exists \phi_M(w) \in Y_M, \; \phi(w) \approx w.$$

It is obvious that the class of mappings $\widetilde{\phi}$ defined as above is the class of surjective mappings.

We assume that if there exists a path $w \in W$ such that $X_w \cap M = \emptyset$, then an empty finite sequence is included in Y_M. Let us introduce the function $\Phi(Y_M) = \sum_{x \in M} \mu(x)$ which defines the total weight of the set M. Then our problem can be rewritten as follows

$$\Phi(Y_M) \xrightarrow[M]{} \min, \quad |Y_M| = |W|. \tag{1}$$

We shall consider a special case of this problem, when all processes in data processing system are homogeneous with respect to the registration complexity. Then, obviously, $\Phi(M) = |M|$ and the problem (1) transforms to the following simpler form

$$|M| \longrightarrow \min, \quad |Y_M| = |W|. \tag{2}$$

The solution of this problem is based on the branch and bound method. We introduce the following notation:

$$Y_m \supset Y_M^{W_i} = \{y \in Y_M : \exists w \in W_i, \ y \approx w\}.$$

The set M *does not identify* the subset W_i of the set of paths W if and only if $|Y_M^{Y_m}| = 1$.

The set M *identifies* the subset W_i of the set of paths W if and only if $|Y_M^{Y_m}| = |W_i|$.

Since it is obvious that $Y_M \equiv Y_M^W$, the condition $|Y_M| = |W|$ of the problems (1) and (2) means that M identifies all the set W, that is

$$\forall w_1 \neq w_2, \ w_i \in W, \ \exists y_1, y_2, \ y_i \in Y_M : \ y_1 \neq y_2, \ y_1 \approx w_1, \ y_2 \approx w_2.$$

The problem (2) is solved in accordance with the branch and bound algorithm by consecutive choosing the "good" numbers-vertices until a set M^* that identifies all the set W is found, which is the solution of the problem. Therefore, consecutively moving over the nodes of the branching tree the elements of the set M are accumulated. In order to choose in each step the needed vertex of the graph G and associated with this vertex the node of the branching tree, one should determine the lower bound of $|M^*|$ and the lower bound for every $x \in X \setminus M$ with respect to the vertex M of the branching tree.

In order to find the lower bounds we have to solve an auxiliary problem. Let A is a finite set, $m = |A|$, and R is the order relation on A. We write elements of A in accordance with the given order R: $A = a_1, a_2, \ldots, a_i, \ldots, a_m$. There exist N^1 subsets of elements of the set A such that the order R is preserved,

$$N^1 = 2^m - 1. \tag{3}$$

Whence, we get the following statements.

Proposition 1. *Let W_M^i be the set of paths unidentifiable by the set M, $0 \leq i \leq 2^{|M|}$. And let p be an integer part of solution of the equation*

$$2^m - 1 = \max_i |W_M^i| \tag{4}$$

with respect to m. Then, in order to identify all the set W, it is necessary to add to the set M not less than p new vertices.

The proof of this proposition is immediately follows from (3).

Let us denote by $\Delta_M = p$ an integer part of solution of the equation (4), that is

$$\Delta_M = \left[\log_2\left(\max_i |W_M^i|\right)\right]. \tag{5}$$

Proposition 2. *If $M_1 \subseteq M_2$, then $W_{M_1}^* \subseteq W_{M_2}^*$, where $W_{M_i}^*$ is corresponding to i the sets of different $M_i, i = 1, 2$, paths of the graph G.*

Corollary 1. *Addition of new vertices to M does not exclude different paths from W_M^*. Whence it follows that the lower bound $n(M)$ of the accumulated set M is equal to $n(M) = |M| + \Delta_M$, and the lower bound of the vertex $x \in X \setminus M$ with respect to M is equal to $n_x(M) = n(M \cup x) = |M| + 1 + \Delta_{M\cup x}$.*

For convenience, we shall write $\Delta_x(M)$ instead of $\Delta_{M\cup x}$. Thus, we get $n_x(M) = n(M \cup x) = 1 + |M| + \Delta_x(M)$ for the lower bound $n_x(M)$ of the vertex $x \in X \setminus M$ with respect to the vertex M of the branching tree, where $\Delta_x(M)$ is computed by the formula (5).

With the use of the obtained expressions, the solution algorithm of the problem is as follows. At the initial step the tree root $M_0 = \emptyset$; according to the Proposition 1, the lower bound of the root equals to $\Delta_M = [\log_2 |W|]$. The rational move over the vertices of the branching tree M_i: in the given step, the vertex of the graph $G(X, E)$ with respect to the given M_i is regarded the best if it has the least lower bound with respect to M_i, that is, the vertex is chosen using the condition

$$n_{x^*}(M_i) = \min_{x \in M} n_x(M_i).$$

In case of equal estimates one chooses the vertex x^*, for which one has

$$\max_j \left|W_{x^*\cup M_i}^j\right| = \min_{x \in X \setminus M} \max_j \left|W_{M\cup x}^j\right|. \tag{6}$$

If we have several vertices satisfying (6), then one can choose any of them or introduce an additional heuristic rule. Since the method of rational move from the root of the tree to the next vertex of branching, and further, is given, hence the procedure for getting the first record is fully defined.

Suppose some admissible solution $M^1 \subset X$ is found as the first record. To find the second record, we consider only those admissible solutions M, for which we have $|M| \leq W/2$. Then one has two situations.

1. Admissible solution is found: $M^2 : |M^2| \leq (M^1)/2$. Hence we found the second record and one searches the third record M^3 such that $|M^3| \leq (M^2)/2$, and so on.
2. Solution is not found. Then, as the second record we consider those admissible solutions M, for which $|M| \leq \left(|M^1| + |M^1|/2\right)/2$, and so on.

The dichotomic search is finished when some record M^* is found, for which $|M^*| \leq |M^i| - \Delta$, where Δ is an indifference threshold, defined by the system designer (developer, administrator), and M^i is the last intermediate record. If such M^* does not exist, then the global extremum is accepted to be M^i. The set M^* is that we sought for, and it identifies all the paths in the graph $G(X, E)$ and uniquely determines the registration graph $R(A, B)$.

5   Conclusions

In this paper we studied the problems of reliability and safety of data processing systems functioning. We have proposed an approach based on the concepts of registration system/archive, registration graph, and processing technology graph. The graph-theoretic model and analysis of reliability and safety of database systems based on the introduced concepts have been presented. The problem has been reduced to the problem of combinatorial optimization on graphs. There are many directions of further research. It would be natural to study the model on graph having inhomogeneous vertices and edges, that is, when the registration costs and expenses of information registration, and comlexities of data processing system procedures are all variable values. The proposed model will be significantly enriched if considered in "random" formulation. We plan to address these issues and study these directions in our future works.

Acknowledgments. This work was supported by the Security Engineering Research Center, granted by the Korea Ministry of Knowledge Economy.

References

1. Singh, S.K.: Database Systems: Concepts, Design and Applications. Dorling Kindersley, New Delhi (2009)
2. Garcia-Molina, H., Ullman, J.D., Widom, J.: Database systems: The Complete Book. Pearson Prentice Hall, Upper Saddle River (2009)
3. Coronel, C., Morris, S., Rob, P.: Database Systems: Design, Implementation, and Management. In: Cengage Learning, Boston (2009)
4. Becker, S.A.: Developing Quality Complex Database Systems: Practices, Techniques, and Technologies. Idea Group Publishing, Hershey (2001)
5. Wang, Y., Li, Z.-h., Xu, J.: Protecting and Recovering Database Systems Continuously. In: Dong, G., Lin, X., Wang, W., Yang, Y., Yu, J.X. (eds.) APWeb/WAIM 2007. LNCS, vol. 4505, pp. 765–776. Springer, Heidelberg (2007)
6. Hosain, M.S., Alam, M.S.: Better Reliability Assessment of Database Based Application Software. In: Proceedings of the 9th International Conference on Circuits ICC 2005, Article No. 54. World Scientific and Engineering Academy and Society (WSEAS), Stevens Point, Wisconsin, USA (2005)
7. Zaniolo, C.: Advanced Database Systems. Morgan Kaufmann, San Francisco (1997)
8. Cooke, R.M.: The Design of Reliability Databases, part I: Review of Standard Design Concepts. Reliability Engineering and System Safety 51, 137–146 (1996)
9. Fragola, J.R.: Reliability and Risk Analysis Database Development: a Historical Perspective. Reliability Engineering and System Safety 51, 125–136 (1996)

10. Bagdasaryan, A.G.: Mathematical and Computer Tools of Discrete Dynamic Modeling and Analysis of Complex Systems in Control Loop. Int. J. Math. Models Methods Appl. Sci. 2, 82–95 (2008)
11. Hadzilacos, V.: An Operational Model for Database System Reliability. In: Proceedings of the 2nd ACM SIGACT-SIGMOD Symposium on Principles of Database Systems PODS 1983, pp. 244–257. ACM, New York (1983)
12. Hadzilacos, V.: A Theory of Reliability in Database Systems. J. ACM. 35, 121–145 (1988)
13. Heimdahl, M.P.E.: Safety and Software Intensive Systems: Challenges Old and New. In: Future of Software Engineering FOSE 2007, pp. 137–152. IEEE Computer Society Press, Washington (2007)
14. Diestel, R.: Graph Theory. Springer, Heidelberg (2006)
15. Gross, J.L., Yellen, J.: Graph Theory and its Applications. Chapman & Hall/CRC (2006)

Mobile Specification Retrieval Methods

Haeng-Kon Kim[1] and Youn-Ky Chung[2,*]

[1] Department of Computer Information & Communication Engineering,
Catholic University of Daegu, Korea
`hangkon@cu.ac.kr`
[2] Department of Computer Engineering, Kyung Il University, Republic of Korea
`ykchung@kiu.ac.kr`

Abstract. A semantic network is a network which represents semantic relations among concepts. This is often used as a form of knowledge representation. This paper introduces a retrieval method based on semantic networks. A hyperlink information is essential to construct semantic networks. The information is very useful as it provides summary and further linkage to construct Semantic networks that has been provided by human. It also has a property which shows review, relation, hierarchy, generality, and visibility. Using this property, we extracted the keywords of mobile documents and made up of the Semantic networks among the keywords sampled from mobile pages. This paper extracts the keywords of mobile pages using anchor text one out of hyperlink information and makes hyperlink of mobile pages abstract as the link relation between keywords of each mobile page. We suggest this useful retrieval method providing querying word extension or domain knowledge by Semantic networks of keywords.

1 Introduction

At present, a retrieval engine is essential to search for the information in the internet. However, it does not present a satisfactory solution to user's desire for the information. Most of retrieval engines store keywords of mobile documents, which are indexed. Retrieval results differ according to the selection and combination of words, because they are presented through matching to user's queried words. User makes efforts of reading many mobile documents to find out one which can satisfy his/her desire for the information among listed mobile documents.

The most frequent words are considered as keywords in keyword extracting method of mobile documents [1]. Keyword extracting method through context analysis has time and technical limit to have to process a natural language. The list of mobile documents which have only user's queried words and keywords in a retrieval result has a limit listing unnecessary mobile documents due to the words of the same, redundant, or various meaning. A retrieval method based on concept has been researched to provide concerned documents or domain knowledge using concept in

---

* Corresponding author.

T.-h. Kim et al. (Eds.): DTA/BSBT 2011, CCIS 258, pp. 71–79, 2011.

order to complement the limit [2, 3, and 4]. Hyperlink is drawn up by author of mobile document, and it is composed of anchor text and link to referred mobile document. When we move to another mobile document from a mobile one, we refer to anchor text. Anchor text is recognized implicitly as the representation of the contents of mobile document connected by hyperlink. Keyword extracting method using an explanation described by author is more accurate and effective than the extraction through context analysis. Anchor text is good information for mobile document. Existing anchor text in mobile document, however, was not connected to the mobile one of hyperlink, but to the mobile one of anchor text [5].

Hyperlink information has used only literal meaning, and it has overlooked the relationship between mobile documents [6, 7].

In this paper, keywords of mobile documents can be extracted quickly using anchor text of hyperlink information. Hyperlink between mobile documents is abstracted into a link between keywords of each mobile one. We present a method which makes possible semantic networks retrieval providing query extension or domain knowledge by construction of semantic networks of keywords.

2 Semantic Networks-Based Retrieval

Semantic networks concept is the process of expanding a particular word to a similar word or a relational one. Semantic networks retrieval is the method which makes possible similar, redundant, and hierarchical representation of word with retrieval expansion using conceptual relationship of words by analyzing word meaning without relying on its spelling only. The method is similar to human way of thought, and it is more effective than other retrieval one. Existing retrieval methods list concerned mobile documents through simple matching queried words to words in mobile documents. In the method, the selection and combination of words is important, because the listed contents are different according to the selection and combination of words. In case of simple matching, no classification of similar and redundant words increases amount of listed documents.

Concept extracting method is divided into a classification of documents in a predefined concept, and the clustering with automatic generation of concept without predefined concept.

INQUERY and EXCITE are the system to classify document using predefined concept. INQUERY system uses the method to map concept to document using modified Bayesian network, after it conceptualizes predefined terminology dictionary using LEX [8]. EXCITE system uses easily extensible "Intelligent Concept Extraction" using modified "Latent Semantic Indexing" without counting on computational ability of computer [4]. The system which generates and extracts a concept automatically without predefined concepts draws up and constructs thesaurus and concept respectively based on the contents of documents using genetic algorithm and self-organized network [5]. The method which is extensible makes an automatic classification, but it requires so much time for initial work. In addition to the method, standardization of IETF is proceeding for concept-extracting work in retrieval system based on additional entry of information for concept extraction by mobile document author with Meta tag added to HTML [6]. The method is so semantic networks in retrieval engine without the necessity of concept-extracting work, and it does not degrade the performance even with increased storage of documents. In case of mobile documents without the information, however, other methods should be used. Concept is used for retrieval expanding inputted query or using various methods to represent concept.

Query informs the system of the fact user intends to retrieve. There are queries of extensive subject, details, and similar documents [8]. Extensive subject query means that it is searched for easily from many documents full of the concerned facts. For example, documents resulting from query like "search for the information about mobile browser" can be represented in various words. Automatic and effective method, however, should be provided, because many concerned documents make accurate results difficult. Detail query is the most of queries made to a retrieval engine. Necessary information of it is contained in a few documents only. If the words are changed, the result is difficult to be searched for. Query of similar documents produces the result by measuring the similarity of document. The query requires natural language process rather than simple word matching. The weakness of details query is that a single query cannot deal with tremendous amount of data processed by each retrieval system. With a single query, threshold of the result is much higher than one of the result determined by the system. Therefore, it should have multiple queries, not a single query. In order to make multiple ones, we should have basic knowledge or clear understanding about the things we want to know.

In this paper, we present a more effective method in processing extension of details query and wide-ranged subject query. Extracted keywords should be detailed for processing details query. This paper uses simple keywords. It also uses information visualization technology for representation of the result.

Information visualization is the technology to help to understand information easily and to support user's decision, by diagrammatizes it after analyzing the information of large scale database.

3   Mobile Specification Retrieval Using Semantic Networks

We propose the method which makes possible semantic networks retrieval in extracted keywords with composition of using association and hierarchy of characteristics of hyperlink, after the keywords of mobile document is extracted quickly and easily by using abstraction of hyperlink information. Hyperlink is drawn up by an author of mobile document. It is composed of a link to refer mobile document and its simple explanation. The method to extract keywords using the explanation described by the author is more accurate and effective than keyword extraction through context analysis. Hyperlink between mobile documents also composes semantic networks of keywords of cach mobile page. Semantic networks introduce information visualization to concept. It is made by connecting words only which have a particular kind of concept. It is diagrammatize with relationship of the words. In this paper, it is used for extending query or providing domain knowledge.

3.1   Concept Extracting Method

This paper uses anchor text and title tag of mobile document for extraction of mobile document keywords. Anchor text is drawn up directly by an author. All the mobile documents have anchor text (generality). Therefore, anchor text can be applied to keyword extraction of all the mobile documents. Title of mobile document is obtained using <Title> tag, and anchor text is obtained in the other mobile document linking (referring to) the mobile one. Anchor text is not drawn up by the author of incited document, but abstracted by an author of inciting one. Its title is extracted directly by the author of incited document. The title is only one, but anchor text is more than one in case many mobile documents are incited. Keywords are extracted with each words weighted. Extracted keywords can be more accurate, because they are based on not mechanical method but anchor text drawn up by author. We design the basic query processing engine as figure 1.

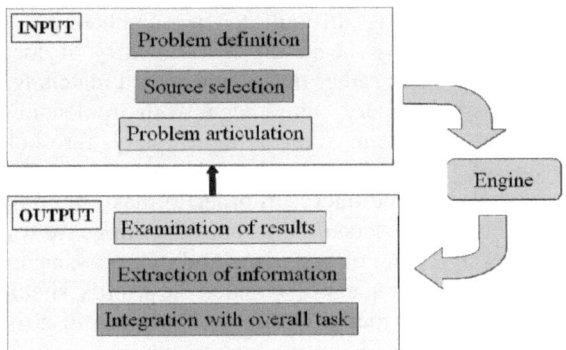

Fig. 1. Basic Query Processing Concepts

3.2 Query Processing

Conceptualized contents are not stored statically, but generated dynamically in execution of query. That is because the construction of semantic networks for all the queries has the restriction of time and space. The method to process a query is that mobile documents having X as a keyword are searched for, after a particular word X is inputted. This document is defined as sibling mobile page. Mobile document which is connected in hyperlink from sibling mobile page is defined as hyper mobile page. The keyword of hyper mobile page is defined as hyper word. Semantic networks are composed of query and the keywords with more accurate concept among hyper keywords. Query processing algorithm is as follows figure 2.

4 System Design

Basically system is composed of spider, indexer, query processor, and visual interface. Spider stores collected mobile pages in database, indexer extracts and indexes keywords based on contents of database, query processor processes query, and visual interface shows result of query in visual effect. [Fig. 3]

1. input Query
2. **If** Query is in index
Collect URLs with query
Store The URLs in murky
Else Return
3. For each URL stored in murky
1. Obtain hyperlink information which URL has
2. Store hyperlink information in m_DestURLKey
4. For each URL stored in m_DestURLKey
1. Obtain keywords which URL has
2. Store keywords in m_listKeyword(hyper keyword list)
5. List sorting by keyword frequency
6. Make string of specified format for representation of semantic networks with words of query and concept
7. Send parameter to Java Applet

Fig. 2. Query Processing Algorithm

A whole system is divided largely into off-line batch job and on-line processing. Off-line batch job constructs index with collected mobile pages, finally produces index file system. On-line processing shows result of semantic networks using index file system with user's query. The processing to collect data is necessary persistently, and it takes so much time because of tremendous amount of data. The processing can be done in batch irrespective of user's response. On the contrary, system response time should be Semantic networks for query processing. Because of that, collection and query processing are separated for off-line and on-line respectively. Its whole operation is as follows: mobile pages are collected in off-line and stored in database, and then index file is generated through indexer. Retrieval system gets query through mobile server. And then query processor accesses to index file through index engine, draws up semantic networks, and provides result of retrieval for user.

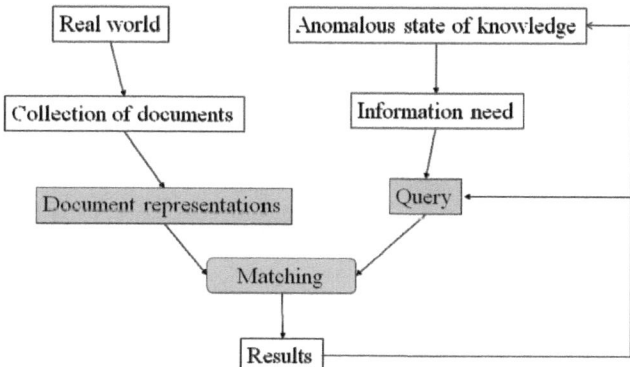

Fig. 3. Formalized Query Process

4.1 Indexer

Indexer is the program which constructs index file system for retrieval service from database composed of mobile page and hyperlink. If SQL statements are used for retrieval service, it takes much time for join and select and much redundant information from database occurs for storage of hyperlink information. A unique index is constructed to reduce redundancy. Its structure is composed of indexes of URL and keyword in order to access hyperlink information quickly and easily. URL Index table plays a role of extracting quickly information for given URL, and it is made of simple hash function. Keyword index table stores information about a list of URLs with given keyword and about the number of given keyword in whole documents.

5   Experiment and Evaluation

In case of query of major concept, semantic networks are composed of the words with equivalent relationship of each other, not with hierarchical relationship. That is, in case of "computer", the words of "science", "university", and "information" are represented [Figure 4]. In case of "science", that internal circle is represented large means that query connects many mobile pages concerned with "science". In case of "information", internal circle is represented relatively small. It means that query has not much relationship with "information", but more relationship with other word.

When it is made of the word to major concept like "computer" in keyword-based retrieval engine, a query is likely to be insignificant with so many results represented. This system, however, helps to select easily detailed concept to be searched for by providing concerned semantic networks. Query of minor concept as a word concerned with detailed concept has a characteristics of low level category among concepts existing in index file. Extracted word as keyword used frequently is not of no value unconditionally. The word of major concept likes "university" appears frequently in semantic networks, because it is connected to many concepts. The word of major concept, however, has no meaning for extension of query. It can be extended to query of college to study "neural network" and college for particular people.

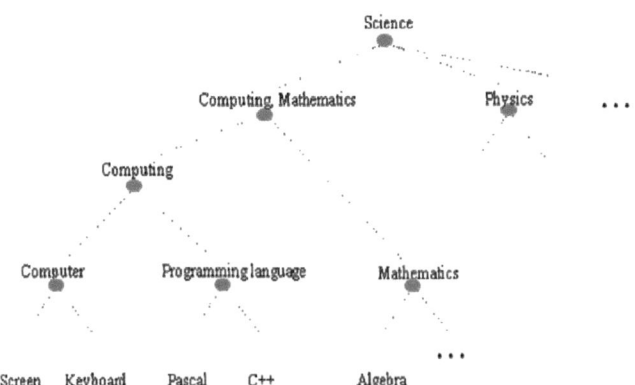

Fig. 4. Example of semantic networks with Major Concept Query

Query syntax is similar to the N/Triples representation:

```
<$1> <http://www.w3.org/1999/02/22-rdf-syntax-ns#type> <http://www.w3.org/2002/07/owl#Ontololology>
<$1> <http://purl.org/dc/elements/1.1/title> <"Ontology">
<$2> <http://www.w3.org/1999/02/22-rdf-syntax-ns#type> <http://www.w3.org/2000/01/rdf-schema#Class>
<$2> <http://www.w3.org/2000/01/rdf-schema#comment> <"*programmer*">
```

This Query contains four conditions which must be met in a matching document:

contains an element $1 which has type #Ontology
contains an element $1 which has title "Ontology"
contains an element $2 which has type #Class Fig.1. Query Processing Algorithm

contains an element $2 which has a comment containing "programmer"

Keywords ("Ontology", "programmer") are extracted from this query and used to query Google in the same way as would be done for the standard search, and any results which come back are downloaded and added to the repository.

Fig. 5. Query Processing Examples

Table 1. Comparison of semantic networks retrieval system

	CE Pro	SemioMap	This system
concept construction	. thesaurus use	. thesaurus use	.hyperlink use
characteristics	. query extension . domain knowledge provided . external retrieval engine	. domain knowledge provided	. query extension . domain knowledge provided . self-retrieval engine
result	. 2D graph	. 3D graph . concerned document	. 2D graph . mobile page list as a result
advantage	. possible construction of only concept of particular field . free query extension	.free search for graph	. concept construction without thesaurus . possible indication of association between concepts . possible keyword extraction without storage of contents of main text . DB saved, speedup
disadvantage	. limit of general-purpose use . processing needed for thesaurus construction	. limit of general-purpose use . processing needed for thesaurus construction	. Hyperlink exists necessarily . concept of meaning-overlapping words extracted . impossible search for multi-stepped graph . query extension of only two steps

And unreachable mobile pages occur among URLs obtained as a result, because mobile pages appear and disappear so semantic networks in mobile. The best advantage of this paper is to generate concept without thesaurus. It takes so much time to construct thesaurus, and method and space for storage are necessary. There is much difficulty in applying promptly all the words generated newly in the area developing rapidly like internet. With thesaurus, however, elaborate concept with no error can be extracted. It requires tremendous time and efforts to make thesaurus as in general-purpose service of large capacity like retrieval system, and it is difficult to apply the thesaurus made like that. That is because thesaurus should be made for the words of the entire field. Therefore, construction of thesaurus for application of concept in retrieval system of general purpose should be automatic, or concept should be extracted in other methods. The method proposed in this paper has the restriction that document should be on mobile site, because concept or keyword cannot be extracted from document without hyperlink. At present, information of Korean language cannot be processed due to analysis of Korean language morpheme. Figure 5 show the query processing examples in our work.

6 Conclusions

In this paper, internet retrieval method is proposed using semantic networks. In this paper, hyperlink information is used to compose semantic networks. Hyperlink information has useful information, because author abstracts and links concerned anchor text in hyperlink. We extract keywords of mobile pages using hyperlink characteristics and compose semantic networks between them. Weighted multiple keywords are extracted using anchor text of abstracted information of hyperlink, and semantic networks is composed by abstraction of the relationship between mobile pages through hyperlink into the relationship between multiple keywords. Keyword extraction method using hyperlink has an advantage to extract simple keywords by using abstraction information drawn up by an author, and experimentally keywords of mobile pages can be searched for relatively exactly. Keyword extraction of much more mobile pages is possible, because it can be extracted by hyperlink information of mobile pages only without processing main text of particular mobile pages for keyword extraction. Semantic networks have an advantage to represent domain knowledge of a real world without thesaurus by representing relationship between keyword nodes. User can extend query without enough knowledge about it. Its disadvantage is that unnecessary information is represented in semantic networks with redundancy of keywords without stemming. Concept cannot be also extracted from documents without hyperlink information. For future work, more effective method which is not simple frequency, and multiple queries processing, need to be researched in selection of concerned words for semantic networks. The method is also necessary for drawing multiple-stepped graph as well as single-stepped graph.

References

1. Preece, J., Rogers, Y., Sharp, H.: Interaction Design: Beyond Human-Computer Interaction. John Wiley and Sons, New York (2002)

2. Jones, M., Marsden, G.: Mobile Interaction Design. Wiley (2006)
3. Agre, P.: Changing Places: Contexts of Awareness in Computing. Human-Computer Interaction 16(2), 177–192 (2001)
4. Dourish, P.: Seeking a foundation for context-aware computing. Human-Computer Interaction 16(2), 229–241 (2001)
5. Dey, A.K., Abowd, G.D., Salber, D.: A conceptual framework and a toolkit for supporting the rapid prototyping of context-aware applications. Human-Computer Interaction 16(2), 97–166 (2001)
6. Graham, C., Kjeldskov, J.: Indexical Representations for Context-Aware Mobile Devices. In: IADIS e-Society, Lisbon, IADIS, Portugal (2003)
7. Morse, D., Armstrong, S., Dey, A.K.: The What, Who, Where, When, Why and How of Context-Awareness. In: CHI. ACM (2000)
8. Kjeldskov, J., Paay, J.: Augmenting the City: The Design of a Context-Aware Mobile Web Site. In: DUX. ACM, San Francisco (2005)
9. Anagnostopoulos, C.B., Tsounis, A., Hadjiefthymiades, S.: Context Awareness in Mobile Computing Environments. Wireless Personal Communications Journal (2006)
10. Henricksen, K., Indulska, J., Rakotonirainy, A.: Modeling Context Information in Pervasive Computing Systems. In: Mattern, F., Naghshineh, M. (eds.) PERVASIVE 2002. LNCS, vol. 2414, pp. 167–180. Springer, Heidelberg (2002)
11. Berry, M.J.A., Linoff, G.: "link analysis".: Data Mining Techniques: For marketing, Sales, and Customer Support, pp. 216–242. Wiley Computer Publishing (1998)
12. Brin, S., Page, L.: The Anatomy of a Large-Scale Hyper textual Mobile Search Engine. In: Proceeding of the Seventh International World Wide Mobile Conference (2002)
13. Callan, J.P., Bruce Croft, W., Harding, S.M.: The INQUERY Retrieval System. In: Database and Expert Systems Applications, pp. 78–83 (1992)
14. Carrie, J., Kazman, R.: Mobile Query: Searching and Visualizing the Mobile through Connectivity. In: Proceeding of the Sixth International World Wide Mobile Conference (1997)
15. Chen, H., Schuffels, C., Orwig, R.: Internet Categorization and search: A Self Organizing Approach. Journal of Visual Communication and Image Representation 7, 88–102 (1996)
16. Klienberg, J.M.: Authoritative Sources in a Hyperlinked Environment. In: Proceedings of the ACM-SIAM Symposium on Discrete Algorithms, pp. 668–677 (1998)
17. Koster, M.: ALIMOBILE: Archie-like indexing in the mobile. Computer Networks and ISDN Systems 27, 175–182 (1994)
18. Marchiori, M.: The Quest for Correct Information on the Mobile: Hyper Search Engines. In: Proceeding of the Sixth International World Wide Mobile Conference (1997)
19. Leavitt, M.: Mobile-agent related research at the CMT. In: Proceedings of the ACM Special Interest Group on Networked Information Discovery and Retrival (1994)

A Comparative Analysis of Array Models for Databases

Peter Baumann and Sönke Holsten

Jacobs University, D-28759 Bremen, Germany
{p.baumann,s.holste}@jacobs-university.de
www.jacobs-university.de

Abstract. While the database collection types set, list, and record have received in-depth attention, the fourth type, array, is still far from being integrated into database modeling. Due to this lack of attention there is only insufficient array support by today's database technology. This is surprising given that large, multi-dimensional arrays have manifold practical applications in earth sciences (such as remote sensing and climate modeling), life sciences (such as microarray data and human brain imagery), and many more areas.

To overcome this, addition of multi-dimensional arrays as a database abstraction have been studied by various groups worldwide. In the attempt towards a consolidation of the field we compare four important array models, AQL, AML, ARRAY ALGEBRA, and RAM. As it turns out, ARRAY ALGEBRA is capable of expressing all other models, and additionally offers functionality not present in the other models. This establishes a common representation suitable for comparison and allows us discussing the commonalities and differences found.

1 Introduction

In 1993, Maier and Vance [18] observed that database technology was rarely used in scientific applications. In their opinion this is due to the lack of support for ordered data structures in database management systems. They showed why it is necessary to give direct support for ordered data structures and presented their ideas on what issues need to be considered when querying ordered data structures, in particular array data.

Applications of the array abstraction are manifold. Generally speaking, arrays occur as sensor, image, and statistics data. In the earth sciences, we find 1-D sensor time series, 2-D satellite imagery, 3-D image time series, and 4-D ocean and atmospheric data. In the life sciences, human brain CAT scan analysis operates on 3-D/4-D imagery, likewise gene expression analysis. Astrophysics, aerodynamic engineering, and high-energy physics comprise further application domains. To the best of our knowledge, no rigid requirements analysis is available currently, only high-level studies like [18] and isolated investigations. For example, the Discrete Fourier Transform (DFT) has been analysed from a database viewpoint [24] and a classification of geographic raster operations from an array query perspective has been published in [10].

In this paper we address array support in databases from a conceptual perspective. The term *array* is seen here in a programming language sense and synonymously to *raster data*, *regularly gridded data*, and *Multi-Dimensional Discrete Data* (MDD) [8].

We consider three algebrae (AML, ARRAY ALGEBRA, RAM) and one calculus (AQL). Approaches can roughly be classified into extended relational models which

T.-h. Kim et al. (Eds.): DTA/BSBT 2011, CCIS 258, pp. 80–89, 2011.

additionally support array data (AQL, RAM) and dedicated array query engines designed for getting embedded into an (R)DBMS (ARRAY ALGEBRA, AML). ARRAY ALGEBRA is unique in that it is implemented in the *rasdaman* array DBMS which is commercialized and in operational use since many years. In *rasdaman*, ARRAY ALGE-BRA defines not only query language semantics, but also storage mapping as well as logical and physical optimization.

There are some more array models in the field which we have not considered in our comparison as they are not as immediately relevant as the candidates chosen. The AQUERY system, which is targeted at financial stock analysis, uses the concept of arrables – i.e., ordered relational tables – and SQL queries extended with an ASSUM-ING ORDER clause [15]. AQuery only supports one-dimensional arrays.

Maier and Howe pursue an ADT/blob based approach where an algebra for the manipulation of irregular topological structures is applied to the natural science domain [11]. As such, it transcends the scope of our analysis.

Cerveiro Cordero at al [5] propose a model which is rather similar to ARRAY ALGE-BRA without sorting. They give, however, an interesting new implementation strategy based on automata. In [9] and [21], modeling of arrays and sequences in Datalog is investigated. SciDB [23] is a system under development which is announced to offer array support; an interesting twist is that ragged arrays are said to become possible. Yet, a formal array model has not been published to the best of our knowledge.

SRAM [6], developed by the originators of RAM [29,1], has a query language under development, but not yet a formal framework. The same holds for SciDB [23] Domain-specific approaches come from the geographic (in particular: remote sensing) field; they include MapScript [25] and a 3-D spatio-temporal extension [14] of Tomlin's map algebra [28]. While there are interesting features and results, we focus on general-purpose, domain-independent models in this contribution.

The remainder of this contribution is structured as follows. The next section introduces and analyzes each of the models selected. Section 3 gives a synoptic comparison. Section 4 presents conclusion and outlook.

2   Overview of Array Models

In this Section, the four array models under consideration are presented. To ease feature comparison, AQL, AML, and RAM are mapped to ARRAY ALGEBRA.

2.1   Array Algebra

For ARRAY ALGEBRA [2,3], image processing and computer graphics [13,7,12] have been studied; AFATL Image Algebra [27] has provided particular insights.

Interval Arithmetics. We assume usual vector notation and operations. In ARRAY ALGEBRA, a *domain* $X \subseteq \mathbb{Z}^d$ of *dimension* $d > 0$ is spanned by two vectors l and h of dimension d as $X := \{p = (p_1, \ldots, p_d) \subseteq \mathbb{Z}^d | \forall 1 \leq i \leq d : l_i \leq p_i \leq h_i\} = [l_1 : h_1, \ldots, l_d : h_d]$. Set X is also referred to as *m-interval* (for *multi-dimensional interval*). The set of all domains is denoted as $\mathbb{D}$.

On such domains, ARRAY ALGEBRA defines some probing functions. Let domain $X = [l_1 : h_1, \ldots, l_d : h_d]$ be given for some $d > 0$. Then, $dim : \mathbb{D} \to \mathbb{N}, dim(X) = d$ is called *dimension of X*, $lo : \mathbb{D} \to \mathbb{Z}, lo(X) = (l_1, \ldots, l_d)$ denotes the low bound corner of X, $hi : \mathbb{D} \to \mathbb{Z}, hi(X) = (h_1, \ldots, h_d)$ denotes the high bound corner of X, and $card : \mathbb{D} \to \mathbb{N}, card(X) = \prod_{i=1}^{d}(hi_i(X) - lo_i(X) + 1)$, or $|X|$, is called *extent of X*.

Sub-array extraction is done through *subsetting*, which we further subdivide into *trimming* and *slicing*. Trimming extracts some subinterval from an m-interval, preserving its dimension. Its formal definition runs as follows. Let X be an m-interval of dimension $d > 0$, spanned by d-dimensional vectors l and h. For some integer i with $1 \le i \le d$ and a one-dimensional interval $I = [m : n]$ with $l_i \le m \le n \le h_i$, the *trim of X to I in dimension i* is defined as

$$trim(X, i, I) := [l_1 : h_1, \ldots, m : n, \ldots, l_d : h_d] = \{x \in X : m \le x_i \le n\}$$

Conversely, *slicing* cuts out a hyperplane, thereby reducing array dimensionality by 1. Formally, for some m-interval X as above, an integer i with $1 \le i \le d$ and an integer s with $lo_i(X) \le s \le hi_i(X)$, the *slice of X at position s in dimension i* is given by
$slice(X, i, s)$
$:= [lo_1(X) : hi_1(X), \ldots, lo_{i-1}(X) : hi_{i-1}(X), lo_{i+1}(X) : hi_{i+1}(X), \ldots, lo_d(X) : hi_d(X)]$
$= \{x \in \mathbb{Z}^{d-1} | x = (x_1, \ldots, x_{i-1}, x_{i+1}, \ldots, x_d), (x_1, \ldots, x_{i-1}, s, x_{i+1}, \ldots, x_d) \in X\}$

The Core Model. Let X be a finite m-interval and V a non-empty set with equality $. = . : V \times V \to boolean$. A *V-valued array A* over domain X is defined as a total function $A : X \to V, A(x) = v$ for $x \in X, v \in V$. We sometimes abbreviate this as $A \in V^X$. The positions $x \in X$ are referred to as *cells*, their associated values $A(x)$ as *cell values*. Let A be a V-valued array over domain X. Then, $dom : V^X \to \mathbb{D}, dom(A) = X$ denotes the *domain of A* while $dim : V^X \to \mathbb{N}, dim(A) = dim(dom(X))$ is the *dimension of A*.

The *array constructor*, MARRAY, establishes an array and initializes its cell values by evaluating some given expression for every cell. Let X be a spatial domain, V a value set, ID be a non-empty set of identifiers, and e_x be an expression with result type V which may contain free occurrences of an identifier $x \in ID$. Then, array A with domain X and cell values e_x for each $x \in X$ is generated as $A = MARRAY(X, x, e_x)$

The *condense operator*, COND, reduces an array to a scalar value by combining the array cell values through some aggregating function. Let o be a commutative and associative operation over V with signature $o : V \times V \to V, x \in ID$ be a free identifier, $X = dom(A) = \{x_1, \ldots, x_n | x_i \in X\}$ an m-interval, and $e_{A,x}$ an expression of result type V possibly containing occurrences of array A and identifier x. Then, the *condense of A by o* is defined as $COND(o, X, x, e_{A,x}) := \bigcirc_{x \in X} e_{A,x} = e_{A,x_1} o \cdots o e_{A,x_n}$.

As an example, for color table computation one has to know the set of all values occurring in the array. The condenser allows to derive this set by performing the union of all cell values: $COND(\cup, dom(A), x, \{A[x]\})$

The third and last core operator is an *array sorter*, SORT, which proceeds along a selected dimension to reorder the corresponding hyperslices. It does so by means of some order-generating expression which allows to rank the slices. The sorted array has the same dimensionality and extent as the original one.

Let A be a d-dimensional array with domain X and value set V, a with $1 \leq a \leq d$ a dimension number, i an index position on dimension a somewhere within A, and r an expression of some type R on which a total ordering $\_ < \_$ is defined and which may contain occurrences of A, a, and i. Let further S be an auxiliary array with $dim(S) = 1$, $dom(S) = [lo_a(dom(A)) : hi_a(dom(A))]$, and values consisting of a permutation of interval $[lo_a(dom(A)) : hi_a(dom(A))]$ resulting from sorting $\{lo_a(dom(A)), \ldots, hi_a(dom(A))\}$ according to the ranking results of expression r on every slice $slice(A, a, i)$. Then, the *array A sorted along dimension a by way of expression s* is given by

$$SORT(A, a, r) = MARRAY(X, x, A(x_1, \ldots, x_{a-1}, S(x_a), x_{a+1}, \ldots, x_d))$$

We observe that, although sorting is defined in terms of slices along a given axis, the ordering predicate does not have to constrain itself to just inspecting this slice. Rather, any general predicate on the array can be phrased, such as evaluating temporal development of values by comparing a slice with its neighbor slices along the time axis.

Derived Operators. As a syntactic convenience we extend the bracket notation on m-intervals so as to allow trimming and slicing of arrays. An *index operation* on a d-dimensional array consists of a bracketed list of d items where each item applies to its dimension sequentially. A pair $l : h$ at position i performs a trim operation in dimension i, while a single item s performs a slicing. For example, expression $A[x_0 : x_1, y, z]$ represents a 1-D array with domain $[x_0 : x_1]$ obtained by trimming A in its first dimension and slicing it in its second and third dimension. Such a combination of trim and slice operations can be rewritten in a natural way using MARRAY together with suitable array addressing arithmetics.

Let T, U, and V be value sets and $f : T \rightarrow V$ and $g : T \times U \rightarrow V$ be unary and binary functions between the value sets. Further, let arrays $A \in T^X$ and $B \in U^X$ be given for some m-interval X. Then, the *induced array operations f and g* are defined as $f : T^X \rightarrow V^X, f(A) = MARRAY(X, x, f(A[x]))$ and $g : T^X \times U^X \rightarrow V^X, g(A, B) = MARRAY(X, x, g(A[x], B[x]))$, resp. Note that ARRAY ALGEBRA does not require any specific cell type function to be induceable. Rather, it defines a generic mechanism which an implementation may or may not offer on particular functions.

Condenser shorthands perform aggregation without explicit cell addressing; hence, they bear resemblance to the relational aggregates. For example, to add up all values in array A over some domain $X \subseteq dom(A)$ a convenience function $add()$ can be defined by $add\_cells(A) = COND(+, X, x, A[x])$. By way of variation, maximum and minimum of the array values can be determined by $max\_cells(A) = COND(max, X, x, A[x])$, $min\_cells(A) = COND(min, X, x, A[x])$. On boolean arrays we can define quantifiers by using the boolean connectors $\wedge$ and $\vee$ to consolidate the values: $all\_cells(A) = COND(\wedge, X, x, A[x])$ and $some\_cells(A) = COND(\vee, X, x, A[x])$. Again, suppressing the iteration variable aids greatly in optimizing.

2.2 AML

AML, short for *Array Manipulation Language*, is an algebra-based, high-level language designed to allow querying array data and defining new arrays in terms of existing ones [19,20]. The model is aiming towards applications in image databases, particularly for remote sensing, but it is described as customizable such that it can serve a wide variety of application domains.

Model. AML uses x for an infinite vector of integers and $x[i]$ to denote its ith element. In AML's terminology an array A is described by a *shape A*, a *domain $\mathscr{D}_A$* and a *mapping $\mathscr{M}_A$*. In passing we note that AML arrays always have a lower bound of 0.

Particular to AML is the notion of bit patterns, which replace indices as a means of accessing arrays within operations. A *bit pattern P* is an infinite binary vector which can be represented in some finite form, i.e. consists of infinite repetitions of some finite vector, such as $P = (1,0)$ which is equivalent to $P = (1,0,1,0,\ldots)$. Along with bit patterns, two *pattern functions* are introduced: *count* and *index*. Function $count(P,k)$ determines the number of zeros in a bit pattern P up to position k while $index(P,k)$ returns the position of the k-th 1 in P.

The three core operators of AML are *subsample*, *merge*, and *apply*. Notably, AML does not provide an array-generating construct "from scratch", like ARRAY ALGEBRA does – new arrays can only be derived from existing ones.

The *subsample* operator is used to eliminate cells in an array following some regular pattern. The *merge* operator combines two arrays. Given two arrays A over domain X and B over domain Y, a dimension number d, a bit pattern P and a default value δi, the merge operation intertwines the arrays along the given dimension according the given pattern filling up holes with the default value. This is written as $MERGE_d(A,B,P,\delta)$.

The *apply* operator serves to apply a given function to an array, similar to the other formalisms.

Mapping to Array Algebra. Any AML array A can be mapped to an ARRAY ALGEBRA array B. We observe that, in general, the spatial domain X is limited to a hypercube in $\mathbb{N}^d$ located at the origin. The array's value set is given by $V := \mathscr{D}_A \cup NULL$, although AML does not make any statement about $\mathscr{D}_A$. We exemplify the mapping by way of the merge operator.

This operator intertwines the elements of two arrays $A \in V^X$ and $B \in V^Y$ along a dimension i according to a bit pattern P, filling up "holes" with a default value $\delta \in V$. Again making use of the pattern functions we can define the ARRAY ALGEBRA domain of the resulting array $C \in V^Z$: $Z := [0 : max(hi_1(X), hi_1(Y)), \ldots, 0 : max(index(P, hi_i(X)), index(P, hi_i(Y))) + 1, \ldots, 0 : max(hi_k(X), hi_k(Y))$].

$S_{P,i}^B : Z \to Y$ can be defined the same way by taking the boolean complement $\overline{P}$ instead of P. As before, "holes" appearing in the merging process, i.e. cells addressed by the step function that are neither located in the domain of A nor in that of B, "holes" are filled with δ values. Now the *merge* function can be written as $MERGE_i(A,B,P,\delta) \equiv MARRAY(Z, x, (P(x(i)) = 1 ? g(A, S_{P,i}^A(x), \delta) : g(B, S_{P,i}^B(x), \delta)))$.

2.3 AQL

AQL is based on NRCA, an extension of the nested relational calculus NRC introduced in [16,17]. Its authors position AQL such as to support the application domains of the NetCDF data exchange format [22].

Model. *NRC* is equipped with complex objects, including products and sets. The value set comprises all complex types mentioned earlier, but plus an uninterpreted base type, i.e., a "black box" with implementation dependent semantics. NRCA mainly adds natural numbers to this model: constants, basic arithmetics, an index set generator, and a summation construct, which allows for expressing aggregates. This allows to algorithmically generate and manipulate arrays.

Operations. NRCA introduces four array constructs, two of which operate on arrays. The first is *subscripting*, written as $e_1[e_2]$ where e_1 is an array and e_2 is an index value; the result is the value contained in the cell addressed thereby. The second is a construct to obtain the *length* in a given dimension d, denoted by $dim_d(e)$. Further, *Array tabulation* (i.e., generation) is done by means of index values and function application on such index values, denoted by $[[e|i_1 < e_1, \dots, i_d < e_d]]$ where $i_1, \dots, i_d$ are the index values and e represents the body of a lambda abstraction $\lambda(i_1, \dots, i_d).e$. The last operator, *index*, converts a set of (index,value) pairs into an array, denoted by $index_d(e)$, where e denotes the set and d refers to the dimensionality of the array to be created. Based on this calculus, AQL is defined using a comprehension syntax that allows for simplified expressions on top of core *NRCA*.

Mapping to Array Algebra. Any array A in AQL can be mapped to an ARRAY ALGEBRA array $B \in V^X$ as defined in subsection 2.1. In general, an AQL domain X is limited to a hypercube in $\mathbb{N}^d$ located at the origin, so it always holds that $lo(X) = 0$. As for the operations, we skip the auxiliary functions (subsetting and length) and instead concentrate on tabulation as the core construct. Mapping to ARRAY ALGEBRA is straightforward: $[[e|i_1 < e_1, \dots, i_d < e_d]] \equiv MARRAY([0:e_1-1, \dots, 0:e_d-1], v, e)$.

The index function is a special case in that it operates on sets of pairs, something not supported by ARRAY ALGEBRA. However, following ARRAY ALGEBRA's philosophy we can assume declarative access operators for data structures provided, in this case: an associative set accessor $ACC : P(I \times V) \times I \to V$ which, for some given set of (index,value) pairs $S \in P(I \times V)$ and a given index value $i = (i_1, \dots, i_d) \in I$ retrieves the corresponding value $v \in V$ such that $ACC(S,i) = v$. With this associative set accessor, the index operation can be phrased as an MARRAY:

$$index_d(e) \equiv MARRAY([0:e_1-1, \dots, 0:e_d-1], x, ACC(e,x))$$

As a side effect, this allows to convert a relational array representation (as used, e.g., in ROLAP) to an array.

2.4 RAM

The RAM model is designed as an extension to the neo-relational DBMS [6].

Model. An array is defined to be a function $A : \mathscr{S}_A \to \tau_A$ where $\mathscr{S}_A$ is the array's *shape* and τ_A is the array's *element type*. The *valence* $|S_A|$ of an array A is defined as the number of dimensions in its shape. An *index value* is a vector $\bar{i} \in S$. Only minor adjustments need to be made to relate RAM's terminology to ARRAY ALGEBRA: The ARRAY ALGEBRA domain is given by an array's shape $\mathscr{S}_A$, its value set by the element type τ_A, and the function itself by A.

Operations. RAM's primary operator is a generic array constructor which reminds of AQL's tabulating construct: $A = [f(i_0, \ldots, i_{(n-1)}) | i_0 < \mathscr{S}_A^0, \ldots, i_{(n-1)} < \mathscr{S}_A^{(n-1)}]$. It specifies an array A of shape S_A with cell values $A(i_0, \ldots, i_{(n-1)}) = f(i_0, \ldots, i_{(n-1)})$ $\forall (i_0, \ldots, i_{(n-1)}) \in \mathscr{S}_A, n = |\mathscr{S}_A|$.

The basis of RAM is its more low-level intermediate algebra. It comes with six basic operators: *const*, *grid*, *map*, *apply*, *choice*, and *aggregate*. These operators can be combined to express an operator equally expressive as the array comprehension above.

The *map* operator creates a new array of which each element is the result of applying a given function to aligned elements in a set of arrays, similar to ARRAY ALGEBRA's unary induced operations: $map(f, A1, \ldots, Ak) := [f(A1(\bar{i}), \ldots, Ak(\bar{i})) | \bar{i} < \mathscr{S}_A]$ where $\mathscr{S}_A = \mathscr{S}_{A1} = \ldots = \mathscr{S}_{Ak}$.

Mapping To Array Algebra. A RAM array A can be mapped to an ARRAY ALGEBRA array $B \in V^X$ as follows. The lower bound of the spatial domain is located at the origin, i.e. $lo(X) = 0$, hence $hi(X) = \mathscr{S}_{\mathscr{A}}$. Furthermore, $V = \tau_A$ and $B(x) = A(X)$.

Mapping the six RAM operators to ARRAY ALGEBRA is straightforward as the *MARRAY* operator can be used to express array comprehensions. For example, the *map* function is represented by an MARRAY expression where function f is applied to the cells of structure $map(f, A_1, \ldots, A_n) \equiv MARRAY([0 : S_1, \ldots, 0 : S_n], v, f(A(v)))$.

3   Comparison

3.1   Array Representation

Domain. All array models introduced share the concept of arrays as functions over a domain, although typing is only supported by ARRAY ALGEBRA. Furthermore, the models agree upon the choice of a rectangular, axis-parallel hypercube in $\mathbb{Z}^d$ as the array's domain. AML, AQL, and RAM constrain array index space to nonnegative values, i.e. the hypercube's lower boundary needs to be located at the origin; ARRAY ALGEBRA allows negative indices as well.

Notational convenience of initial segments of $\mathbb{N}_0$ values and a lack of gain in expressiveness when adding negative indices are RAM's arguments [1]. In [16], lifting this domain restriction is considered to be a valuable extension to AQL as it would allow for more meaningful indices for scientific arrays.

Value Set. RAM, with its purpose being the array component of the MonetDB system, assumes relational attribute types; support for complex types is explicitly omitted for reasons of simplicity. AQL's embedding into the nested relational calculus allows for handling other complex types such as sets and tuples efficiently in the core language.

AML and AQL support nesting, i.e., arrays are allowed as array values. In AML, however, definition of the value set is not concretized at all, leaving value semantics completely to the implementation. This has severe implications on optimizability as the optimizer will run into black boxes while analyzing the query and, hence, cannot natively understand expressions in their entirety. ARRAY ALGEBRA provides a plug-in semantics where arbitrary data types can be accommodated, however, with clear rules on how to orchestrate them into the overall model. Among these explicitly stated requirements are commutativity and associativity of a condenser summarization function (to allow efficiency-increasing rewriting) and the homogenous algebra property (to obtain well-defined induced operators). We, therefore, adopt the position that a concise value set semantics is an asset in any kind of formal array model.

3.2   Operations

Finding a suitable operator set is a major challenge in array models. In most models – ARRAY ALGEBRA, AQL and RAM – a generic "array generator" is defined to possibly cover a large extent of these high-level operators. Although different in style (ARRAY ALGEBRA uses array tabulation while AQL and RAM use array comprehensions) we believe these operators to be of equal expressiveness. AML does not have any "from scratch" constructor. ARRAY ALGEBRA, AQL, and RAM also introduce an explicit aggregation mechanism. ARRAY ALGEBRA additionally proposes a sorting operator.

Expressiveness. Arrays can be represented straightforward as sets of (index,value) pairs; hence, a comparison of array model expressiveness with that of the relational model is a natural question. Libkin, Machlin, and Wong show that NRCA and, hence, AQL have the same expressive power as relational calculus plus ranking [16]. Machlin extends this with important results with regard to the complexity of array indexing; see [16] and, in particular, [17]. ARRAY ALGEBRA adds the *SORT* operator. Intuitively, this means an additional step in expressiveness, but how does this get manifest? We investigate into this by looking at the mapping of the *SORT* operator to the relational model where $SORT(A,1,r)$ can be written as (*select x from R order by* $r(A,x)$) × (*select v from R order by x*). The difference to NRCA and other formalisms is that the sorting criterion, r, is not predefined through some property (such as the total ordering of the cell type) but variable: a user can specify any sorting predicate expressible within the overall framework. In other words, *SORT* actually is a higher-order construct, a functional. Now that we are sensitized we observe that this occurs in another place as well: ARRAY ALGEBRA's $COND(\circ,X,x,e)$ is a higher-order construct which is parametrized with the aggregation operation $\circ$. Conversely, conventional formalisms assume a fixed, hardwired set of aggregations. Situation is similar with *MARRAY*.

This modelling of arrays as second-order formalism seems natural, considering the definition of arrays as being functions. Libkin et al use relational algebra plus ranking for their equivalence proof; however, the ranking property assumes a totally ordered index set and, as such, is only used for establishing the array model and not for introducing an array sorter. As it stands, beyond ARRAY ALGEBRA we are not aware of any formalism containing a sorter.

4   Conclusion and Outlook

In this paper, we have surveyed array database theory which is gradually entering into a consolidation phase. Three main contributions towards this are made in this paper. First, we have presented an overview of four important array database models, thereby discussing commonalities and differences. Further, we have shown that ARRAY AL-GEBRA can express each of these (while the inverse does not hold) by inspecting all relevant aspects of both data model and operations. Finally, discussion of architectural and optimization issues has shown suitability of ARRAY ALGEBRA to support all these levels, up to an implementation, *rasdaman*, which is in successful operational use.

While each of the models has its individual merits and has sound formal arguments for its operator choice, major differences can be found in the operation set chosen and the rigor applied in their semantics definition. Aggregates are seen as important, but sometimes modeled explicitly and sometimes only implicitly. In summary, ARRAY AL-GEBRA is powerful enough to express all models investigated; the inverse is not true, as no other model offers an equivalent to the *SORT* operation. Hence, we feel confident that ARRAY ALGEBRA represents the state of the art in array database modeling.

All in all, albeit young as a database discipline, arrays are making their way to a first-class data abstraction, thereby completing the family of collection types supported by databases. Still, there are manifold research issues in this young discipline. We work on extending the framework beyond arrays towards general meshes so as to allow retrieval on further spatiotemporal scientific data, such as Voronoi-type structures (adaptive grids can be handled already). Use of the *rasdaman* system in further projects (and standardization) in earth, space, and life sciences is expected to unveil new use cases requiring additional functionality and optimizations.

References

1. Ballegooij, A.V., Vries, A.P.D., Kersten, M.: Ram: Array processing over a relational dbms (2003)
2. Baumann, P.: On the management of multi-dimensional discrete data. VLDB Journal Special Issue on Spatial Database Systems 4(3), 401–444 (1994)
3. Baumann, P.: A Database Array Algebra for Spatio-Temporal Data and Beyond. In: Tsur, S. (ed.) NGITS 1999. LNCS, vol. 1649, pp. 76–93. Springer, Heidelberg (1999)
4. Catell, R., Cattell, R.G.G.: The Object Data Standard, 3.0 edn. (2000)
5. Cordeiro, J.P.C., Camara, G., de Freitas, U.M., Almeida, F.: Yet another map algebra. Geoinformatica 13, 183–202 (2009)
6. Cornacchia, R., Heman, S., Zukowski, M., de Vries, A., Boncz, P.: Flexible and efficient ir using array databases. Technical Report INS-E0701 (2007)
7. Felger, W., Frühauf, M., Göbel, M., Gnatz, R., Hofmann, G.: Towards a reference model for scientific visualization systems. In: Proc. Eurographics Workshop on Visualization in Scientific Computing (April 1990)
8. Furtado, P., Baumann, P.: Storage of multidimensional arrays based on arbitrary tiling. In: Proceedings of the 15th International Conference on Data Engineering, March 23-26, pp. 328–336. IEEE Computer Society (1999)
9. Greco, S., Palopoli, L., Spadafora, E.: Extending datalog with arrays. Data Knowl. Eng. 17(1), 31–57 (1995)

10. Gutierrez, A.G., Baumann, P.: Modeling fundamental geo-raster operations with array algebra. In: Workshops Proceedings of the 7th IEEE International Conference on Data Mining, ICDM 2007, pp. 607–612. IEEE Computer Society (2007)
11. Howe, B., Maier, D.: Algebraic manipulation of scientific datasets. In: Proc. VLDB 2004, pp. 924–935 (2004)
12. ISO, editor. Information technology: Computer graphics and image processing, image processing and interchange, functional specification. Part 2: Programmer's imaging kernel system: Application program interface. Number ISO/IEC JTC1 SC24 Document IM-157. International Organization for Standardization, ISO (1992)
13. ISO, editor. Information Processing Systems - Computer Graphics - Computer Graphics Reference Model. Number ISO/IEC JTC1 / SC24 / WG1 N133. International Organization for Standardization (ISO) (August 1990)
14. Mennis, C.T.J., Viger, R.: Cubic map algebra functions for spatio-temporal analysis. Cartography and Geographic Information Systems 30(1), 17–30 (2005)
15. Lerner, A., Shasha, D.: Aquery: Query language for ordered data, optimization techniques, and experiments. In: VLDB 2003, pp. 345–356 (2003)
16. Libkin, L., Machlin, R., Wong, L.: A query language for multidimensional arrays: Design, implementation, and optimization techniques, pp. 228–239 (1996)
17. Machlin, R.: Index-based multidimensional array queries: safety and equivalence. In: Libkin, L. (ed.) PODS, pp. 175–184. ACM (2007)
18. Maier, D., Vance, B.: A call to order. In: PODS 1993: Proceedings of the Twelfth ACM SIGACT-SIGMOD-SIGART Symposium on Principles of Database Systems, pp. 1–16. ACM, New York (1993)
19. Marathe, A.P., Salem, K.: A language for manipulating arrays. In: Proc. of VLDB, pp. 46–55 (1997)
20. Marathe, A.P., Salem, K.: Query processing techniques for arrays. In: SIGMOD 1999: Proceedings of the 1999 ACM SIGMOD International Conference on Management of Data, pp. 323–334. ACM, New York (1999)
21. Mecca, G., Bonner, A.J.: Sequences, datalog and transducers. In: PODS 1995: Proceedings of the Fourteenth ACM SIGACT-SIGMOD-SIGART Symposium on Principles of Database Systems, pp. 23–35. ACM, New York (1995)
22. n.n, http://www.unidata.ucar.edu/software/netcdf (last seen, September 2011)
23. n.n. Scidb, http://www.scidb.org (last seen, September 2011)
24. Buneman, P.: The Fast Fourier Transform as a Database Query. Technical Report MS-CIS-93-37, University of Pennsylvania (1993)
25. Pullar, D.: Mapscript: A map algebra programming language incorporating neighborhood analysis. Geoinformatica 5-2, 145–163 (2001)
26. Ritsch, R.: Optimization and Evaluation of Array Queries in Database Management Systems. Phd thesis, TU Muenchen (1999)
27. Ritter, G., Wilson, J., Davidson, J.: Image algebra: An overview. Computer Vision, Graphics, and Image Processing 49(1), 297–336 (1994)
28. Tomlin, D.: Geographic Information Systems and Cartographic Modeling. Prentice-Hall, Englewood Cliffs (1990)
29. van Ballegooij, A.R.: RAM: A Multidimensional Array DBMS. In: Lindner, W., Fischer, F., Türker, C., Tzitzikas, Y., Vakali, A.I. (eds.) EDBT 2004. LNCS, vol. 3268, pp. 154–165. Springer, Heidelberg (2004)

Potentials of Circulation Data Analysis for Library Marketing: A Case Study in a University Library

Toshiro Minami

Kyushu Institute of Information Sciences, 6-3-1 Saifu, Dazaifu, Fukuoka 818-0117 Japan &
Kyushu University Library, 6-10-1 Hakozaki, Higashi, Fukuoka 812-8581 Japan
minami@kiis.ac.jp

Abstract. Marketing is an essential tool for customer relation management of companies. Contrastingly, non-profit organizations like libraries have not been considered it very useful. However, due to the progress of the information society, our society becomes too complex to capture the patrons's needs for them in the way they have been using so far. The aim of this paper is to demonstrate the importance and usefulness of marketing by presenting some analysis methods of circulation data of a university library and showing what can get about the patrons' behavior. The analysis methods presented in this paper are not only some statistical ones but also non-statistical ones, such as the one for analyzing patron's borrow-return behavior pattern; in other words patron profile. Even though library marketing is in its very early stage, it has a big potential for aiding library management. We have to accumulate the case studies and establish useful marketing methodologies.

Keywords: Data Analysis, Data Mining, Library Marketing, User/Patron Profiling, Circulation Data.

1 Introduction

Mission of library is to provide its patrons (users) with services about information, knowledge, learning, etc. In order to provide with better services, it is important to capture the patrons' needs exactly. However patrons' needs are changing too quickly to capture the needs accurately and keep the library services appropriately. This is a big issue in library marketing. Marketing [1] is crucial not only for libraries but also for all organizations. Especially it is recognized as a live-or-die issue for profit-oriented companies. They put big effort to capturing customers' exact needs [8].

Marketing with data analysis is an essential tool for these companies. It is well known that convenience stores utilize POS data for marketing. E-commerce businesses collect the customers' purchasing data in real-time and use them for letting the customers buy more. The point-card system provides the companies with purchase data and the companies provide the point-card holders with discounts.

In this paper we take the approach of applying such methodology to library marketing. Actually libraries already have a lot of data that are usable in marketing, including circulation (borrow and return) records, patrons' profile information, Web homepage access log data, institutional repository access log data, and others. In this paper we show some example data analysis experiences as a case study using the circulation records of the Central Library of Kyushu University, Japan (KUL) for the academic year 2007. See [5] and [6] for previous works of this paper.

Statistical circulation data analysis is a standard method for libraries in collection evaluation [4], which is useful to capture representative values of the whole data. The system WorldCat Collection Analysis [7], for example, provides an easy-to-use statistical analysis tools to librarians. Yamada analyzed circulation data by considering the material age of books [9]. A research on evaluating the usage of e-books is reported in [4]. As a non-statistical approach, investigating the association rules in classification category of books using a data mining method is reported in [3].

Our approach in this paper is different from such standard methods. We take two different approaches; one is statistical one for surveying the general characters of patrons, and another one is non-statistical one for different types of patron profiling.

In the rest of this paper, we firstly analyze the circulation data of KUL and get the overall patrons' profile in Section 2. Then in Section 3, we put focus on an undergraduate student and try to capture the behavioral profile of the student. Lastly in Sectin 4, we summarize the discussions in this paper.

2   Investigation to Patron Profiling from Circulation Data

We use the circulation data taken in the Central Library of Kyushu University, Japan, of the academic year 2007 (from April 2007 to March 2008). In this section we investigate the overall patron (students, academic staff, administrative staff, etc.) profiles (borrow-return behavior). Then we put focus on the undergraduate students who occupy about half of the records. Lastly we take up a student as a target patron and investigate in more detail about his or her behavior pattern. We are expecting to extract useful tips for library management from such analyses in the near future.

2.1 Data Overview

The circulation data for the academic year 2007 of KUL include 67,304 records. A record item consists of the book ID, classification code, call number, borrower ID, borrower affiliation, borrower type (undergraduate student, masters student, Ph.D. student, professor, staff, others), and timestamps (date: "yyyy:mm:dd" and time: "hh:mm:ss") for borrowing and returning, etc. KUL opened 348 days out of 366 days in this year by considering at least one borrowing should have been recorded in every opening day. Thus about 193 books were borrowed per day in average.

The records say 6,118 patrons borrowed at least one book during this year period. Note that the registered patron who visited the library just for browsing the books and/or for studying and did not borrow even a book was not counted in this number. Of course the patrons who had not visited the library were not counted, neither. If we calculate under such understanding, the average number of circulations per patron is about 11.

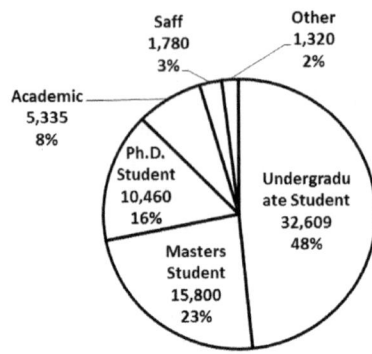

Fig. 1. Ratio of Number of Borrowed Books to Patron Type

Fig. 1 shows the ratio of circulation record numbers classified in the patron types. It says that 48% (32,609 records) among 67,304 circulation records come from the undergraduate students, which consist of 29,698 records by the regular students that belong to the faculties and 2,911 records by other types of students such as non-degree students. Further, master's students occupy 23% (15,800) and Ph.D. students occupy 16% (10,460), and other types of patrons, 13% (8,435).

An entrance data say that about 60% of patrons are undergraduate students. So these records give another evidence that majority patrons are (undergraduate) students for university library. So we may conclude that the university libraries should put more efforts to the improvement of the library services that target the students.

2.2 Analysis of Undergraduate Students' Records

From now on we put focus on the undergraduate students; so we will simply use "students" for "undergraduate students." There were 29,698 records for students. There were 2,966 students borrowed at least one book during the year, thus 10 books were borrowed per student in average. There were 11,822 students in all and 8,921 of them were affiliated in the Hakozaki campus where the central library is located. Thus 25% of students (one out of four) and 33% of students (one out of three) in Hakozaki campus borrowed at least one book. Considering all students in Kyushu University, a student borrowed 2.5 books, and 3.3 books by students in Hakozaki campus.

Public library in Japan often uses the number of books circulated per residence in a year (circulation number divided by population) as an important index for library's performance. More than 10 is a kind of goal for them. It is interesting that this magical number 10 appears in a university library data even if it is just a coincidence.

The total number of books a student borrowed during the year, varies from 208 as the maximum to 1 as minimum. By sorting the numbers of borrowed books per student in descending order from the maximum, they are 208, 181, 172, 143, and 141. Each number is the one borrowed by one student. The number of students who borrowed more than 100 books per year is only 12 (0.4% of students). On the other hand, the minimum number is 1, of course, and the number of students who borrowed only one book in this year is 468; which is 16% of students (among those who borrowed at least one book in the year), that is about one student out of 6.

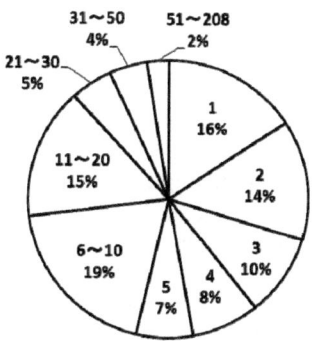 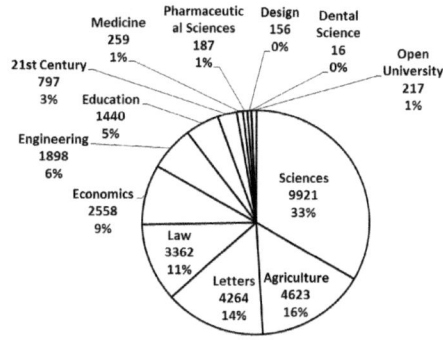

Fig. 2. Ratio of Number of Students to the **Fig. 3.** Ratio to Faculties that Borrower Number of Borrowed Books   Students belongs to

Fig. 2 shows the ratio of numbers of students to the numbers of borrowed books per year. One book is borrowed by 16% (468) of students, followed with 2 books by 14% (413), 3 books by 10% (283), 4 books by 8% (233), and 5 books by 7% (196), and so on. More than half of students who borrowed at least one book, borrowed up to 5 books in a year period! In addition to it, the ratio of the students who borrowed more than 10 books (Note that 10 is the average number of the borrowed books by a student in the year.) is 27% (800 students); about one student out of 4.

Fig. 3 shows the ratio of circulation numbers classified in the borrower's affiliated faculties. We can see that (Faculty of) Sciences occupies 33% (9,921 records), which is the maximum one, followed by Agriculture's 16% (4,623), Letter's 14% (4,264), Law's 11% (3,362), and Economics's 9% (2,558). These faculties are located close to the library so that students can visit there easier than the students of other faculties. This might be a good explanation why these faculties occupy such high ratios than others. However the numbers are affected not only by the distances from the library but also by the numbers of students belonging to the faculties and other factors, we have to invesigate further in order to understand the results with sufficient confidence.

Fig. 4 shows the ratios in accordance with the years of students. The numbers of the circulated books by the 5th and 6th year students are very small; 65 (0.22%) and

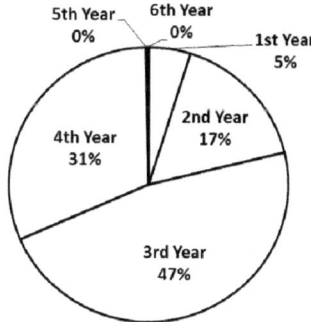

Fig. 4. Ratio to Number of Student Years

50 (0.17%), respectively, so they are marked 0%, which comes from the fact that these students belong to either Medical or Dental Science. In the records for 1st to 4th year students, the number increases from 1st to 3rd, which probably means that the students get familiar with the campus life and thus with their libraries as well. We might be able to say also that the learning subjects are getting more and more difficult as they go from 1st to 3rd year so they get to visit library more often.

The reason why the number for the 4th year is less than the number for the 3rd year, may be that even they need more books and academic papers and documents for their research and writing class assignments, it might be easier to get such technical books and materials either at their campus libraries, faculty libraries, or at their research labs than to get them in the central library. We have to investigate more on this issue.

Fig. 5 shows how the numbers of cirulation varies in a year. Roughly, the first semester in an academic year in Japan starts in April and ends in early August. The second semester starts in October and ends in early February. During the first semester, the number increases monotonically, probably because of the orientations to the libraries that inform the students how to use the libraries. Also the students might start thinking about that they need help from the libraries as they start getting lectures and find some difficulties in learning. The numbers are very small during the summer vacation, i.e. August and September. Among them that for August is bigger than that for September, probably because students still come to campus for lectures and end of semester examinations during early August.

During the second semester, i.e. from October to January, the number is as big as the one for July. However we can see some difference from the first semester. In the second semester the numbers are mostly same during the first three months, probably because the students are sufficiently familiar with using libraries in the second semester. It is interesting to see that the number becomes the highest in January, which probably means it is the closing month for the second semester and the students start thinking about the term-end examinations and they start studying for preparing them. Another reason may include that many students who are supposed to graduate in March have to write up the graduation thesis so they need to survey the books, journals, and papers that are provided by the library.

The numbers decrease a lot in February and March. Just like the comparison between August and September, the number for Frebruary is much bigger than that of March because some people still need to visit libraries because of the similar reason in January. By comparing the numbers in September and March, the number is bigger

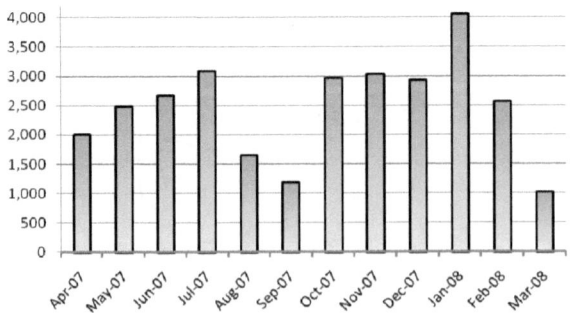

Fig. 5. Number of Circulated Books According to Month of the Year

in September than in March, probably because March is the end of the academic year and thus the most graduating students have to prepare for the life after graduation, so they have little time to keep visiting the libraries.

3   A Student's Behavior Analysis for Patron Profiling

In this section, we take a student as a sample and analyze his or her circulation data intending to find some behavioral features. By accumulating such analysis experiences we would be able to understand the students, and it will help the librarians to create better patron services, which is the eventual goal for the library.

3.1   Choosing the Target Student

We choose the student who borrowed the maximum number of books as the sample student. We will call the student by A. Student A borrowed 208 books, which means that the student borrowed about 0.7 books per the library's opening day. The number of books borrowed by student A is 86. So, a book is borrowed 2.4 times in average.

Student A was affiliated in the Faculty of Sciences, which is located very close to the central library. So, it is very easy for the students to visit the library and student A did visit the library quite often. He or she was in the 4th year in 2007. Student A visited 96 days, which is 27% of the opening days; in other word about 2 days per week. The average number of books borrowed per visiting day is 2.2.

Fig. 6 shows the ratio of the classified field of the borrowed books according to the NDC (Nippon Decimal Classification) system. Physics (NDC 420) occupies the maximum number 139 (67%); much more than half. So we can easily guess that student A affiliated to Department of Physics. After physics books, follows mathematics (NDC 410) with 53 (25%) and these two fields occupy 91%. From this observation, we can guess student A studied theoretical physics very hard. Other subjects student A studied were astronomy, which counts 13 (6%), so he or she seems to have been interested in this subject probably as an application field of theoretical physics. For the books in computer and information sciences, student A borrowed two books, which are handbooks for a graph drawing software GNUPLOT.

Fig. 7 shows the ratio of the number of books borrowed by student A according to the day of week. Student A borrowed the books mostly on Monday, which occupies

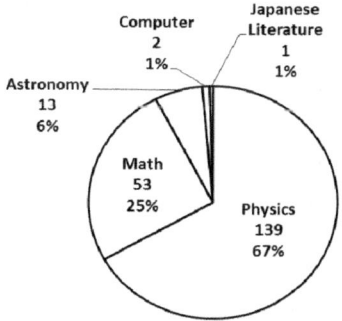

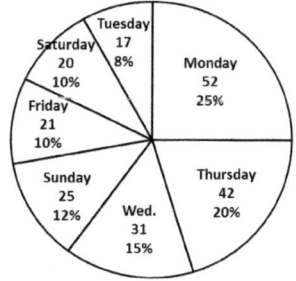

Fig. 6. Ratio to Number of Circulated Books according to Subjects

Fig. 7. Number of Circulated Books According to Day of the Week of Student A

about 25% (1 out of 4 books), followed by Thursday (20%), Wednesday (15%), and the total ratio of these three days is 60%. The next one is on Sunday, when as big as 12% of books are borrowed, even though it is a weekend day. Student A was in the 4th year, and thus he or she was taking possibly just a small number or no classes this year so that he or she can visit library every day. The number is a little bit smaller on Saturday than Sunday (8% to 12%), maybe because student A reserved Saturday for something special such as for housework or something. The minimum number is on Tuesday (8%), which may mean that the student had some other jobs on this day; such as a seminar, or seminars, in the lab or something like that. Anyway, we can feel the student A's life rhythm from these data.

Let us compare the usage pattern of student A with that of all students. Roughly speaking, the usage frequencies for week days are almost same for all students. In weekends, each frequency is about one third of a week day. From these observations we can see that some students visit on specific days of week only and such tendencies differ from student to student. Such biases are canceling each other, and the ratios for the days of the weekday as a whole become almost same. Many students would not visit library during weekends so the ratios for Saturday and Sunday are much lower than other days. By considering other aspects also, we can guess that student A visited the library exceptionally often, because he or she studied very hard, his or her affiliation was the Faculty of Sciences, and thus it was very easy to visit frequently.

Fig. 8 shows the frequencies of the duration of borrowing days of student A. The peak comes at the 14 days because it is the due period days. Its frequency is 74 (35%), which means about one third books were returned on the due days. It is interesting to see that books are returned one week later 8 times (4%) only. This probably means that student A visited the library quite often so that he or she did not think to return the books one week after. Student A returned a lot on the 12th and 13th days, which are the 2 and 1 day(s) before the due day. This means student A wanted to return the books before the due day. On the other hand 32 cases (15%) are returned beyond 14 days, which looks like returning after the due days. However the 6 cases (3%) from 17 to 22 days are borrowed in December and returned in January, so these books are borrowed in the specially extended borrowing periods so they do not mean they were due to the student's mistakes. Thus the 26 cases (13%), one case out of 8, from 15 and 16 days are the one student A actually returned the books after the due days.

Let us compare the frequencies of cases according to the borrowing period of days for all students with that of student A's borrow-return pattern. In this case also, the

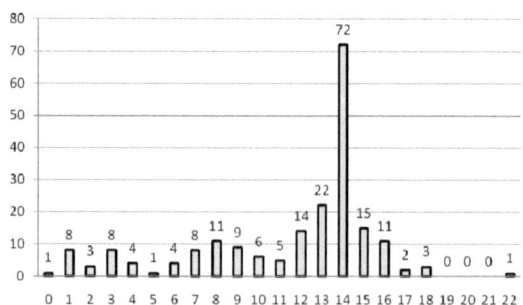

Fig. 8. Frequencies of the Borrowing Days Duration of Student A

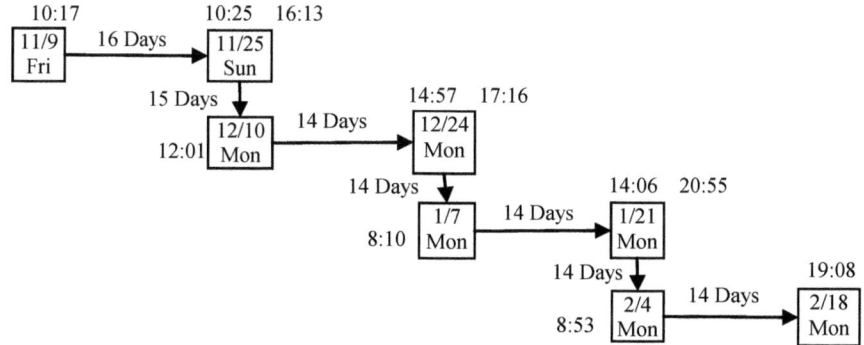

Fig. 9. Borrow-Return Pattern for "Fundamental Electromagnetism I, II"

peak of frequency lies on the 14th day, where about 14% of books are returned on this due day. This ratio is as half as much of that of student A's 35%. The next biggest is the 14th day, one day before the due day, and the 7th day, one week after borrowing. The ratios are 8% and 6%, respectively.

3.2 Borrow-Return Behavior Analysis

The maximum borrowing times of a book by student A was 14. As we see more precisely, we found that "the book" actually consists of a set of two books; namely "Fundamental Electromagnetism I" and "Fundamental Electromagnetism II". These books were borrowed 7 times at the same time as a pair as is shown in Fig. 9. They were firstly borrowed on Nov. 9th (Friday) at 10:17 and returned on Nov. 25th (Sunday) at 10:25; so they were borrowed for 16 days. Then they were borrowed again on the same day, i.e. Nov. 25th, at 16:30 and were returned on Dec. 10th (Monday) at 12:01. They were borrowed on the same day at the same time and returned on Dec. 24. In the last record, they were borrowed on Feb. 4 (Monday) and returned on Feb. 18 (Monday).

We can see also the borrowing days for the first and second times were 16 and 15, respectively. Student A borrowed the books for the next time on the same day even though the student returned the books after due days. In the library's rule, if a patron returned late then the patron is suspended from borrowing during the same length of days as the days of delay. It means even with the late returns of the student, the librarians did not give him such penalties; it is rather preferable to applying the rule in a strict way. It is important for the library to encourage the students to utilize the library as much as possible and study hard. By not applying the rule, student A was able to keep borrowing and studying these books. It is interesting to see that the student returned on the 14th day in the rest 5 times. This fact may indicate that student A started taking more attention to the due days after his or her mistakes at the beginning.

It is also interesting to see that student A borrowed the books for the next time just a short time after he or she returned the books on the first day of each month. It probably means that student A did not have enough time to keep staying in the library on the early days of these months. This guess seems to be appropriate from the time data also; the returning and borrowing times are mostly early in the morning in these cases. On the other hand, the differences between returning time and re-borrowing

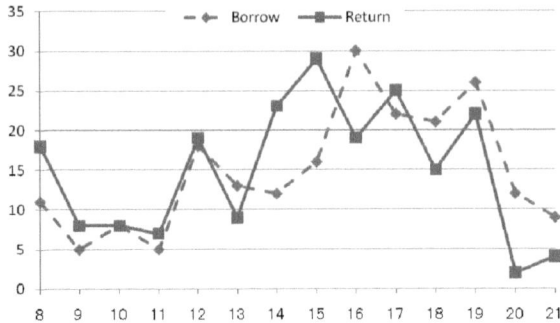

Fig. 10. Distribution Pattern of Borrow and Return of Books according to Time of the Day

time were very long for the second times of the same month; about 5 hours in average. From these records student A seemed to stay at the library for a long time on the later days of these months.

3.3 Time Zone Analysis

Fig. 10 shows the distribution of the frequencies of circulation records of student A over time of a day. We can see easily that student A visited library from early in the morning till late at night, just like he or she visited library constantly on every day of the week. Among them the time zone the student visited most frequentely is from 3 p.m. to 7 p.m., or late afternoon after lunch time.

Let us assume, based on the so far observations about student A's behavior, that he or she returned books just after visiting library and borrowed books just before he or she left library. Then we can say, the time zone when more books were returned than borrowed is the one student A entered library and the time zone when more books were borrowed than returned is the one the student left library.

This assumption seems to be true because the early time zone in the morning, from 8 a.m. to 10 a.m., more books were returned than were borrowed, and late time zone at night, from 6 p.m. toward 10 p.m., more books were borrowed than were returned. We can see also that the time zone from 2 to 3 p.m. is the one for visiting to library. Also we can see that the time close to 5 p.m. is another time for visiting. On the other hand, the time zones around 1 p.m. and 4 p.m. seem to be the leaving times. The former might be for taking lunch and the latter might be either for dinner or for going back home or for something like evening seminars.

From these observations, we can guess student A visited library from the beginning of the service time oftern. He or she left library for just a short time for lunch and dinner. High time of using the library was after lunch time to evening or maybe sometime till night. From this analysis also, student A must have studied very hard and he or she was the heavy library user.

4 Concluding Remarks

In this paper we presented some analysis methods on a circulation data of a library. First, we gave an overview of patrons of the library and pointed out that nearly 90%

circulation records came from students; about half came from undergraduate students. This fact indicates the importance of students for university libraries. Then we put focus on the undergraduate student data. We found the fact that more than half of students borrowed only 5 or less books. So we can say that one of the biggest problems is to encourage the students who visit library to visit more often. By analyzing the affiliation of the students, the distance from the library seems to affect the number of visits. So another problem to be solved is to find a good way so that even the students away from the library, wants to visit, or to introduce a system so that library books and librarians visit the students in order to let them closer.

Then we chose a student who borrowed more books than any other student and had a close look about the student's behavior or lifestyle that can be seen from the circulation data. The student seemed to study physics, probably theoretical physics, and he or she studies hard, visited library quite often not only in the week days but also during weekends. By analyzing how the student borrowed the book, actually a pair of books, which was borrowed more often than other books by the student, the student borrowed the book on the same day he or she returned. In the time zone analysis, the student visited the library from early in the morning till late at night. So we guessed the student might have stayed a long time in the library and studied hard.

The approach presented in this paper is a case study of library data analysis for library marketing. The research and practice for library marketing in the same approach is not so much popular so far. However, by considering the potential usefulness of data analysis which has been proved in other fields like Web marketing and others, such an approach to library marketing should be recognized as an essential one in the future; hopefully in the near future. We have to keep accumulating as many case studies of library data analysis at the moment. Then we can prove its usefulness for library management and improving the patron services.

References

1. American Marketing Association (AMA): Definition of Marketing, http://www.marketingpower.com/AboutAMA/Pages/DefinitionofMarketing.aspx
2. Brooks, F.: The Mythical Man-Month. Addison-Wesley (1995)
3. Cunningham, S.J., Frank, E.: Market basket analysis of library circulation data. In: Proceeding of 6th International Conference on Neural Information Processing, pp. 825–830. IEEE Computer Society, Perth (1999)
4. Littman, J., Connaway, L.S.: A Circulation Analysis of Print Books and e-Books in an Academic Research Library. Library Resources & Technical Services 48(4), 256–262 (2004)
5. Minami, T., Kim, E.: Data Analysis Methods for Library Marketing. In: Lee, Y.-h., Kim, T.-h., Fang, W.-c., Ślęzak, D. (eds.) FGIT 2009. LNCS, vol. 5899, pp. 26–33. Springer, Heidelberg (2009)
6. Minami, T.: Challenge toward Patron Understanding – A Search for Patron's Profile through Circulation Data of Library –, Kyushu University Library: Annual Report 2010/2011, pp. 9–18 (2011) (in Japanese)
7. Online Computer Library Center, Inc (OCLC): WorldCat Collection Analysis, http://www.oclc.org/collectionanalysis/
8. Silverstein, B.: Business-to-Business Internet Marketing. Maximum Press (1999)
9. Yamada, S.: Analysis of Library Book Circulation Data: Turnover of Open-shelf Books. Journal of College and University Libraries 69, 27–33 (2003) (in Japanese)

Minimal Cost Attribute Reduction
through Backtracking

Fan Min and William Zhu

Lab of Granular Computing,
Zhangzhou Normal University, Zhangzhou 363000, China
minfanphd@163.com, williamfengzhu@gmail.com

Abstract. Test costs and misclassification costs are two most important types in cost-sensitive learning. In decision systems with both costs, there is a tradeoff between them while building a classifier. Generally, with more attributes selected and more information available, the test cost increases, and the misclassification cost decreases. We shall deliberately select an attribute subset such that the total cost is minimal. Existing decision tree approaches deal with this issue from a local perspective. They benefit from immediately available test results, therefore objects falling into different branches may experience different tests. In this paper, we consider the situation where tests have delayed results. Since we need to choose a test set for all objects, the attribute reduction problem is defined from a global perspective. We propose a backtrack algorithm with three pruning techniques to find a minimal cost reduct. Experimental results indicate that the pruning techniques are effective, and the algorithm is efficient on a medium sized dataset Mushroom.

Keywords: Cost-sensitive learning, attribute reduction, test cost, misclassification cost, backtrack algorithm.

1 Introduction

In cost-sensitive learning, test costs and misclassification costs [4] are more often addressed. Test cost is the money, time, or other resources we pay for collecting a data item of an object [4,16,7]. When test costs are involved, an attribute value of an object is available only if a test is undertaken. This is why an attribute (or a feature) is also called a *test*. In this context, we would like to select an attribute subset so as to minimize the test cost, and at the same time, preserve a particular property of the decision system. This problem is called the cheapest reduct problem [15], or the minimal test cost reduct problem [6]. The property to be preserved might be the positive region [6,10], conditional information entropy [13], consistency [2], covering [21,20], etc. In case that costs for all tests are the same, this problem coincides with the traditional reduct problem [10,12,17].

Misclassification cost is the penalty we receive while deciding that an object belongs to class J when its real class is K [4,3]. It can be represented by a matrix $C_{k \times k}$, where k is the number of distinct classes. Most research works require $c_{i,i} = 0$ where $0 \leq i \leq k$ since the object is correctly classified (see, e.g., [16]). While others do not have such

T.-h. Kim et al. (Eds.): DTA/BSBT 2011, CCIS 258, pp. 100–107, 2011.

a requirement (see, e.g., [18]). This is due to subsequent operations which turn out to have a certain cost. For example, if a patient is correctly classified as got a flu, there will be a treatment cost. When misclassification costs are considered, we would like to construct a classifier so as to minimize the average misclassification cost [19]. The average misclassification cost is a more general metric than the accuracy. If values on the diagonal of C are identical and values off the diagonal are identical, minimizing the former is exactly equivalent to maximizing the latter [16].

The attribute reduction problem becomes more interesting when we consider both types of costs. With more attributes selected, the test cost increases. At the same time, with more information available, the misclassification cost decreases, or at least does not increase on the training set. Unlike traditional reduct problems, we do not require any particular property of the decision system to be preserved. The objective of attribute reduction is solely to minimize the average total cost. In fact, Turney [16] and Ling et al. [5] have considered this issue. They focused on tests producing immediate results, and employed the C4.5 decision tree approach [11] to select attributes independently in branches. Since tests undertaken in different branches may be different, the test selection issue is addressed from a local perspective.

In this paper, we consider the situation where tests have delayed results. We need to choose a test set for all objects. In other words, we adopt the parallel test assumption [6], and define the problem from a global perspective. We propose a backtrack algorithm to find a minimal cost reduct. Three pruning techniques are employed and discussed. The algorithm is implemented in our open source software Coser [9]. We undertook experiments using the Mushroom dataset from the UCI library [1]. Experimental results show that the pruning techniques are astonishingly effective. Therefore we suggest to employ this algorithm directly in medium sized datasets.

Section 2 defines the minimal cost reduct problem by considering the average total cost. Then Section 3 presents the algorithm along with the analysis of pruning techniques. Experimental settings and results are discussed in Section 4. Finally, Section 5 concludes and suggests further research trends.

2   Problem Definition

In this section, we define the minimal cost reduct problem. Specifically, we present the data model on which the problem is defined, and propose the new problem.

We consider decision systems with test costs and misclassification costs. Since the paper is the first step toward this issue, we only consider simple situations.

Definition 1. *A decision system with test costs and misclassification costs (DS-TM) S is the 7-tuple:*

$$S = (U, C, d, V, I, tc, mc), \tag{1}$$

where U is a finite set of objects called the universe, C is the set of conditional attributes, d is the decision attribute, $V = \{V_a | a \in C \cup \{d\}\}$ where V_a is the set of values for each $a \in C \cup \{d\}$, $I = \{I_a | a \in C \cup \{d\}\}$ where $I_a : U \rightarrow V_a$ is an information function for each $a \in C \cup \{d\}$, $tc : C \rightarrow \mathbb{R}^+ \cup \{0\}$ is the test cost function, and $mc : k \times k \rightarrow \mathbb{R}^+ \cup \{0\}$ is the misclassification cost function, where $\mathbb{R}^+$ is the set of positive real numbers, and $k = |I_d|$.

Terms *conditional attribute*, *attribute* and *test* are already employed in the literature, and these have the same meaning throughout this paper. $U, C, d, \{V_a\}$ and $\{I_a\}$
can be displayed in a classical decision table. Table 1 is an example where $U = \{x_1, x_2, x_3, x_4, x_5, x_6, x_7\}$, $C = \{$Headache, Temperature, Lymphocyte, Leukocyte,
Eosinophil, Heartbeat$\}$, and $d = $ Flu.

Table 1. An exemplary decision table

Patient	Headache	Temperature	Lymphocyte	Leukocyte	Eosinophil	Heartbeat	Flu
x_1	yes	high	high	high	high	normal	yes
x_2	yes	high	normal	high	high	abnormal	yes
x_3	yes	high	high	high	normal	abnormal	yes
x_4	no	high	normal	normal	normal	normal	no
x_5	yes	normal	normal	low	high	abnormal	no
x_6	yes	normal	low	high	normal	abnormal	no
x_7	yes	low	low	low	high	normal	yes

The test cost function is stored in a vector $tc = [tc(a_1), tc(a_2), \ldots, tc(a_{|C|})]$. We
adopt the test cost independent model [7] to define the cost of a test set. That is, the test
cost of $B \subseteq C$ is given by

$$tc(B) = \sum_{a \in B} tc(a). \tag{2}$$

The misclassification cost function can be represented by a matrix $mc = \{mc_{k \times k}\}$,
where $mc_{i,j}$ is the cost of classifying an object of the i-th class into the j-th class. Similar to Turney's work [16], we require that $m_{i,i} \equiv 0$. However, this requirement could
be easily loosed if necessary. The following example gives us intuitive understanding.

Example 1. A DS-TM is given by Table 1, $tc = [2, 2, 15, 20, 20, 10]$, and

$$mc = \begin{bmatrix} 0 & 800 \\ 200 & 0 \end{bmatrix}.$$

That is, the test costs of headache, temperature, lymphocyte, leukocyte, eosinophil, and
heartbeat are \$2, \$2, \$15, \$20, \$20, and \$10 respectively. If a person with (without) a
flu is misclassified as without (with) a flu, a penalty of \$800 (\$200) is paid.

The average misclassification cost is computed on the training set as follows. Let
B be the selected attribute set. $IND(B)$ is a partition of U. Given $U' \in IND(B)$,
$U' \subseteq POS_B(\{d\})$ if and only if $d(x) = d(y)$ for any $x, y \in U'$. In other words, we can
generate a certain rule to classify all elements in U' correctly, and the misclassification
cost of U' is $mc(U', B) = 0$. If, however, there exists $x, y \in U'$ st. $d(x) \neq d(y)$, we
may guess one class for all elements in U'. In fact, there are a number of approaches
[14] for this issue. For simplicity, we always generate the rule which minimizes the total
misclassification cost over U. In this way, the average misclassification cost is given by

$$\overline{mc}(U, B) = \frac{\sum_{U' \in IND(B)} mc(U', B)}{|C|}. \tag{3}$$

Algorithm 1. A backtrack algorithm to a minimal cost reduct

Input: $S = (U, C, D, V, I, tc, mc)$
Output: R and cmc, they are global variables
Method: bamor

1: $R = \emptyset$;//A minimal cost reduct, a global variable
2: $cmc = \overline{c}(U, \emptyset)$;//Currently minimal cost, a global variable
3: $B = \emptyset$;//Currently selected attribute subset
4: backtrack(B, 0);

Because the test cost of any object is the same, the average total cost is given by

$$\overline{c}(U, B) = tc(B) + \overline{mc}(U, B). \tag{4}$$

Now we propose a new definition as follows.

Definition 2. *Let* $S = (U, C, d, V, I, tc, mc)$ *be a DS-TM. Any* $B \subseteq C$ *is called a minimal average total cost reduct iff*

$$\overline{c}(U, B) = \min\{\overline{c}(U, B')|B' \subseteq C\}. \tag{5}$$

Under the context of DS-TM, a minimal average total cost reduct is called a minimal cost reduct for brevity. Naturally, we can define a problem with the objective of searching a minimal cost reduct. This problem will be called the minimal average total cost reduct problem, or the minimal cost reduct problem (MCR). We note the following:

1. The MCR problem has a similar motivation with the cheapest cost problem [15], or the minimal test cost reduct (MTR) problem [6]. Their difference lies in the type of costs considered.
2. When the misclassification cost is too large compared with test costs, the MCR problem coincides with the MTR problem.
3. Although we still call the attribute subset a *reduct*, it is quite different from its traditional counterparts. It does not preserve any property (e.g., positive region [10], information entropy [13]) of the decision system. Nor does it have a constraint [8]. The selection of attributes only relies on the cost information.

3 The Minimal Cost Backtrack Algorithm

This section presents the algorithms. It contains two methods. The first method, as listed in Algorithm 1, initializes and invokes the backtrack algorithm. The second method, as listed in Algorithm 2, deals with the problem recursively.

Generally, the solution space of an attribute reduction algorithm is $2^{|C|}$. Therefore pruning techniques are critical for the backtrack algorithm. There are essentially three pruning techniques employed in Algorithm 2. First, Line 1 indicates that the variable i starts from il instead of 0. Whenever we move forward (see Line 13), the lower bound is increased. For example, if $B = \{a_0, a_3\}$, B' can never be $\{a_0, a_3, a_2\}$. This is because

---

Algorithm 2. The backtrack algorithm

Input: B, il (attribute index lower bound)
Output: R and cmc, they are global variables
Method: backtrack

```
 1: for (i = il; i < |C|; i ++) do
 2:     B' = B ∪ {aᵢ};
 3:     if (tc(B') ≥ cmc) then
 4:         continue;//Prune for too expensive test cost
 5:     end if
 6:     if (c̄(U, B') ≥ c̄(U, B)) then
 7:         continue;//Prune for non-decreasing cost
 8:     end if
 9:     if (c̄(U, B') < cmc) then
10:         cmc = c̄(U, B');//Update the minimal cost
11:         R = B';//Update the minimal cost reduct
12:     end if
13:     backtrack(B', i + 1)
14: end for
```

---

that $\{a_0, a_2, a_3\} = \{a_0, a_3, a_2\}$ might either be checked in an earlier stage, or in a pruned subtree. Without this technique, the solution space could be $|C|!$ instead of $2^{|C|}$.

Second, Lines 3 through 5 indicate if the test cost of B is no less than the currently minimal cost (cmc), there is no need to check B. This is because the misclassification cost is non-negative. Although simple, this technique can prune most branches and make the algorithm applicable for medium sized data (e.g., Mushroom).

Third, Lines 6 through 8 indicate that if more tests produce high cost, the current branch will never produce the total cost reduct. This technique is not as effective as the second one, however it still saves the run time.

4 Experiments

In this section, we try to answer the following questions by experimentation.

1. Is the algorithm efficient?
2. How does the minimal total cost reduct changes for different cost settings?

We choose the Mushroom dataset from the UCI Repository of Machine Learning Databases [1]. It has 22 attributes and 8124 objects. Therefore if we can deal with it in reasonable time, we can also apply the algorithm to many real world applications. We simply let test costs uniformly distributed in [1, 10] with integer values.

We undertake three sets of experiments. The first set studies the change of run time with the misclassification cost. Misclassification costs are identical for different misclassifications. Fig. 1 shows the backtrack steps of the algorithm. Two approaches are compared. The Fast approach is given by the algorithm. While the Slow approach does not employ the third pruning technique. From this figure we observe that

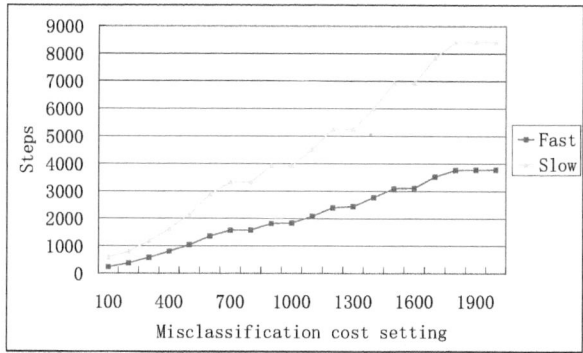

Fig. 1. Backtrack steps

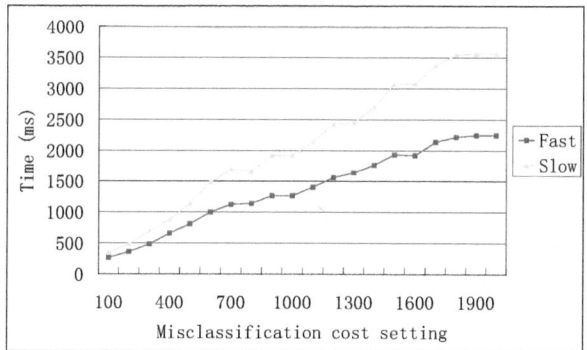

Fig. 2. Run time

1. The second pruning technique is very effective. The number of steps is decreased from 4 million to a few thousand.
2. The third pruning technique is also useful. The number of steps is reduced to about a half further.
3. The number of steps increases with the increase of misclassification costs. This is due to the fact that more tests tend to be checked.

Fig. 2 shows the run time. We know that the time is only a few seconds, which is acceptable for a real world application.

Fig. 3 shows the change of test costs vs. the average total cost. When misclassification costs are low, we would like to avoid undertaking expensive tests. With the increase of misclassification costs, the test cost increases and even equals to the average total cost. This is because the penalty of misclassification is too high, and we would like to obtain more information. In case where the test cost is equal to the average total cost, the MCR problem coincides with the minimal test cost reduct (MTR) problem [6].

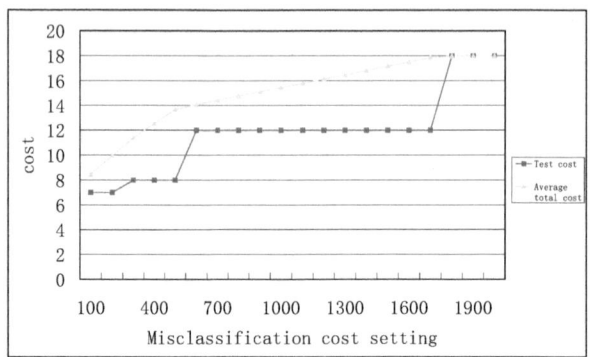

Fig. 3. Test cost and average total cost

5   Conclusions and Further Works

In this paper, we considered the minimal cost reduct problem in decision systems where both test costs and misclassification costs exist. The objective is to minimize the average total cost. The new problem is more general than the minimal test cost reduct problem [6], which is more general than the traditional reduct problem [10,12]. In fact, when misclassification costs are too high compared with test costs, the new problem coincides with the minimal test cost reduct (MTR) problem [6]. We proposed a backtrack algorithm with three pruning techniques to deal with this problem. Experimental results showed that these techniques are effective. The algorithm only takes a few seconds on the Mushroom dataset with 22 tests and 8124 objects. Therefore this algorithm is a good choice for medium sized datasets. It is also useful in evaluating the performances of heuristic algorithms, which should be designed in our further works for large datasets.

Acknowledgements. This work is in part supported by National Science Foundation of China under Grant No. 60873077, 61170128, the Natural Science Foundation of Fujian Province, China under Grant No. 2011J01374 and the Education Department of Fujian Province under Grant No. JA11176. We would like to thank Dominik Ślęzak for his valuable comments.

References

1. Blake, C.L., Merz, C.J.: UCI repository of machine learning databases (1998), http://www.ics.uci.edu/~mlearn/mlrepository.html
2. Dash, M., Liu, H.: Consistency-based search in feature selection. Artificial Intelligence 151, 155–176 (2003)
3. Elkan, C.: The foundations of cost-sensitive learning. In: IJCAI, pp. 973–978 (2001)
4. Hunt, E.B., Marin, J., Stone, P.J. (eds.): Experiments in induction. Academic Press, New York (1966)
5. Ling, C.X., Sheng, V.S., Yang, Q.: Test strategies for cost-sensitive decision trees. IEEE Transactions on Knowledge and Data Engineering 18(8), 1055–1067 (2006)

6. Min, F., He, H., Qian, Y., Zhu, W.: Test-cost-sensitive attribute reduction. Information Sciences 181, 4928–4942 (2011)
7. Min, F., Liu, Q.: A hierarchical model for test-cost-sensitive decision systems. Information Sciences 179, 2442–2452 (2009)
8. Min, F., Zhu, W.: Attribute reduction with test cost constraint. Journal of Electronic Science and Technology of China 9(2), 97–102 (2011)
9. Min, F., Zhu, W., Zhao, H., Pan, G.: Coser: Cost-senstive rough sets (2011), http://grc.fjzs.edu.cn/~fmin/coser/
10. Pawlak, Z.: Rough sets. International Journal of Computer and Information Sciences 11, 341–356 (1982)
11. Quinlan, J.R. (ed.): C4.5 Programs for Machine Learning. Morgan kaufmann Publisher, San Francisco (1993)
12. Skowron, A., Rauszer, C.: The discernibility matrics and functions in information systems. In: Intelligent Decision Support (1992)
13. Ślęzak, D.: Approximate entropy reducts. Fundamenta Informaticae 53(3-4), 365–390 (2002)
14. Ślęzak, D., Widz, S.: Is It Important Which Rough-Set-Based Classifier Extraction and Voting Criteria Are Applied Together? In: Szczuka, M., Kryszkiewicz, M., Ramanna, S., Jensen, R., Hu, Q. (eds.) RSCTC 2010. LNCS, vol. 6086, pp. 187–196. Springer, Heidelberg (2010)
15. Susmaga, R.: Computation of Minimal Cost Reducts. In: Raś, Z.W., Skowron, A. (eds.) ISMIS 1999. LNCS, vol. 1609, pp. 448–456. Springer, Heidelberg (1999)
16. Turney, P.D.: Cost-sensitive classification: Empirical evaluation of a hybrid genetic decision tree induction algorithm. Journal of Artificial Intelligence Research 2, 369–409 (1995)
17. Wang, G.: Attribute Core of Decision Table. In: Alpigini, J.J., Peters, J.F., Skowron, A., Zhong, N. (eds.) RSCTC 2002. LNCS (LNAI), vol. 2475, pp. 213–217. Springer, Heidelberg (2002)
18. Yao, Y., Zhao, Y.: Attribute reduction in decision-theoretic rough set models. Information Sciences 178(17), 3356–3373 (2008)
19. Zhou, Z., Liu, X.: Training cost-sensitive neural networks with methods addressing the class imbalance problem. IEEE Transactions on Knowledge and Data Engineering 18(1), 63–77 (2006)
20. Zhu, W.: Generalized rough sets based on relations. Information Sciences 177(22), 4997–5011 (2007)
21. Zhu, W., Wang, F.: Reduction and axiomization of covering generalized rough sets. Information Sciences 152(1), 217–230 (2003)

Protein Function Prediction by Spectral Clustering of Protein Interaction Network

Kire Trivodaliev, Ivana Cingovska, and Slobodan Kalajdziski

Ss. Cyril and Methodius University, Faculty of Computer Science and Engineering,
Ruger Boskovic 16, 1000 Skopje, Macedonia
{kire.trivodaliev,ivana.cingovska,
slobodan.kalajdziski}@finki.ukim.mk

Abstract. The increasing availability of large-scale protein-protein interaction (PPI) data has made it possible to understand the basic components and organization of cell machinery from the network level. Many studies have shown that clustering protein interaction network (PIN) is an effective approach for identifying protein complexes or functional modules. A significant number of proteins in such PIN remain uncharacterized and predicting their function remains a major challenge in system biology. We propose a protein annotation method based on spectral clustering, which first transforms the PIN using the normalized Laplacian of the PIN graph, and then employs a classic clustering algorithm like k-means. Protein functions are assigned based on cluster information. Experiments were performed on PPI data from the bakers' yeast and since the network is noisy and still incomplete, we use pre-processing and purifying. We also performed network weighting based on the annotation correlation between nodes. Results reveal improvement over previous techniques.

Keywords: Protein interaction networks, Protein function prediction, Spectral clustering.

1 Introduction

Within cells, proteins seldom act as single isolated units to perform their functions. It has been observed that proteins involved in the same cellular processes often interact with each other [1]. Protein-protein interactions are thus fundamental to almost all biological processes [2]. With the advancement of the high-throughput technologies, such as yeast-two-hybrid, mass spectrometry, and protein chip technologies, huge data sets of protein-protein interactions become available [3]. Such protein-protein interaction data can be naturally represented in the form of networks, which not only give us the initial global picture of protein interactions on a genomic scale but also help us understand the basic components and organization of cell machinery from the network level.

An important challenge for system biology is to understand the relationship between the organization of a network and its function. It has been shown that clustering protein interaction networks is an effective approach to achieve this goal

T.-h. Kim et al. (Eds.): DTA/BSBT 2011, CCIS 258, pp. 108–117, 2011.
© Springer-Verlag Berlin Heidelberg 2011

[4]. Clustering in protein interaction networks is to group the proteins into sets (clusters) which demonstrate greater similarity among proteins in the same cluster than in different clusters. Since biological functions can be carried out by particular groups of genes and proteins, dividing networks into naturally grouped parts (clusters or communities) is an essential way to investigate some relationships between the function and topology of networks or to reveal hidden knowledge behind them.

Classical graph-based agglomerative methods employ a variety of similarity measures between nodes to partition PPI networks, but they often result in a poor clustering arrangement that contains one or a few giant core clusters with many tiny ones [5]. To improve the clustering results, PPI networks were weighted based on topological properties such as shortest path length [6,7], clustering coefficients [8], node degree, or the degree of experimental validity [9]. As a new type of clustering algorithms, the edge-betweenness was defined as a global measure to separate PPI networks into subgraphs in a divisive manner [10-12]. Edge-betweenness is the number of shortest paths between all pairs of nodes that run through the edge. It is able to identify biologically significant modular structures, but it requires lots of computational resources. As an approach to coordination of typical clustering algorithms, an ensemble method was proposed to combine multiple, independent clustering arrangements to deduce a single consensus cluster structure [13].

Not only network partition but also extraction of protein complexes have been performed to analyze PPI networks. To detect such densely connected subgraphs in them, many algorithms were proposed. Molecular Complex Detection (MCODE) is based on node weighting by local neighborhood density and outward traversal from a locally dense seed protein to isolate densely connected regions [14]. The Restricted Neighborhood Search Clustering Algorithm (RNSC) is a cost based local search algorithm to explore the solution space to minimize cost function, calculated according to the numbers of intra-cluster and inter-cluster edges [15]. In [16] Physical model-based algorithms were presented, such as Monte Carlo optimization (MC) and Superparamagnetic Clustering (SPC), to identify densely connected regions. MC formulates the finding of highly connected as an optimization problem: find a set of n vertices that maximizes the density of the subgraph. SPC is a hierarchical clustering algorithm inspired from an analogy with the physical properties of a ferromagnetic model subject to fluctuation at nonzero temperature. MC performs better for clusters that share common vertices and for high density graphs, whereas SPC has an advantage identifying clusters that have very few connections to the rest of the graph.

Heuristic rule-based algorithms were proposed to reveal the structure of PPI networks [17,18]. A layered clustering algorithm was presented, which groups proteins by the similarity of their direct neighborhoods to identify locally significant proteins that links different clusters, called mediators [19]. Power graph analysis transforms biological networks into a compact, less redundant representation, exploiting the abundance of cliques and bicliques as elementary topological motifs [20]. On the other hand, interestingly, spectral clustering analysis, which is an appealing simple and theoretically sound method [20,21], has hardly been studied to partition PPI networks, while it is used for detecting protein complexes [16].

The quality of the obtained clusters can be evaluated in couple of ways. One of the criteria rates the clustering as good if the proteins in a cluster are densely connected between themselves, but sparsely connected with the proteins in the rest of the network [22]. Some systems provide tools for generation of graphs with known clusters, which is modelled with the parameters of the explored network [23]. Then the clusters obtained with the clustering algorithm are compared to the known ones. The clustering method can also be evaluated by its ability to reconstruct the experimentally and biologically confirmed protein complexes or functional modules [12][16][22].

In this paper we set up a framework for predicting protein function by using clustering in PIN. We use spectral clustering, which first transforms the PIN using the normalized Laplacian of the PIN graph, and then employs a classic clustering algorithm like k-means. Protein functions are assigned based on cluster information. Experiments were performed on PPI data from the bakers' yeast and since the network is noisy and still incomplete, we use pre-processing and purifying. We also performed network weighting based on the annotation correlation between nodes.

2 Research Methods

The methods for protein function prediction by clustering of PIN generally consist of three phases, as represented on figure 1.

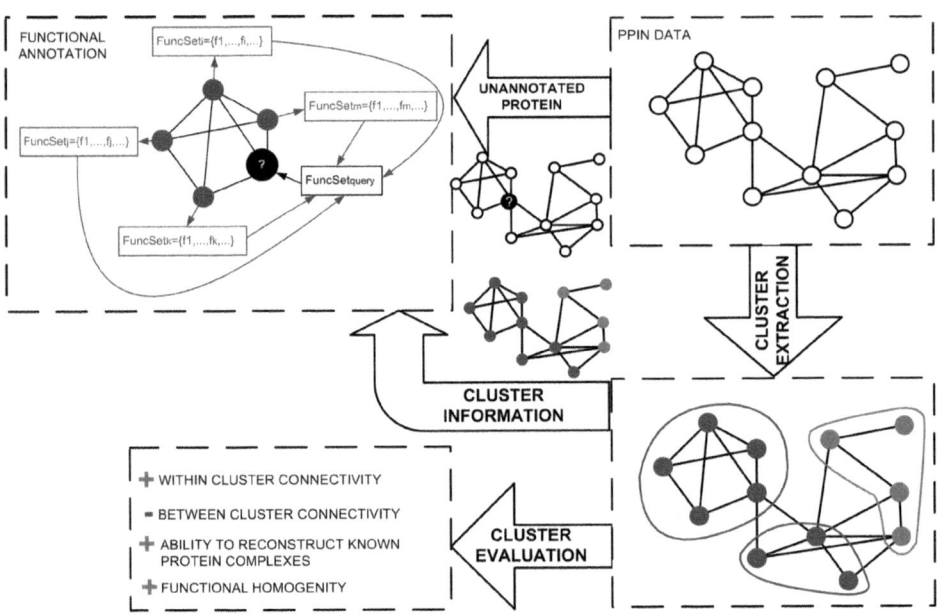

Fig. 1. General framework for protein function prediction by clustering in PIN

The first phase is dividing the network in clusters, using its topology or some other information for the nodes or the edges, if such an information is available. The compactness and the characteristics of the obtained clusters are then evaluated in the second phase. From physical aspect the clusters can be assessed by the ratio of the number of edges within and between the clusters, and from biological aspect they can be assessed by the functional and biological similarities of the proteins in the clusters. This second phase is not mandatory, but might be useful because it can point out what to expect from the function prediction itself. The prediction of the protein annotations for the proteins in the clusters is the task of the third phase.

2.1 Cluster Extraction Based on Spectral Analysis of the Protein-Protein Interaction Graph

One of the basic types of graph clustering is the spectral clustering, which performs spectral analysis of the graph's adjacency matrix or some of its derivatives, by finding its eigenvalues and eigenvectors [26]. The first step in the spectral clustering is transforming the initial dataset into a set of points in an n-dimensional space, whose coordinates are elements of n selected eigenvectors. This change in the representation of the data enhances the characteristics of the clusters making them more distinctive. Then a classical clustering algorithm, like k-means for example, can be used, to cluster the data.

Let $G = (V;E)$ be the graph representing a protein-protein interaction network, where V is the set of nodes (proteins), and E is the set of unweighted/weighted undirected edges, and A be the column normalized adjacency matrix defined by E. Although the initial idea for spectral analysis was intended directly to the adjacency matrix of the graph, the newer algorithms use the Laplacian matrix L, which is derived from the adjacency matrix A as in equation (1).

$$L = D - A . \tag{1}$$

In this equation, D is a diagonal matrix whose diagonal element D_{ii} equals the degree of the node i of the graph. We than compute the first k generalized eigenvectors $u_1,..., u_k$ of the generalized eigenproblem

$$Lu = \lambda Du . \tag{2}$$

Equation (2) is also referred to as normalization of the Laplacian matrix.

The main characteristic of the graphs' Laplacian matrix is the fact that the number k of zero eigenvalues equals to the number of connected components of the graph. The non-zero values of the corresponding eigenvectors are on the indices of the nodes that belong to the corresponding connected component. If those eigenvectors are put as columns of one $U^{|V|xk}$ matrix, each row represents one node which has only one non-zero value: on the position of the eigenvector of the connected component it belongs to. If the graph consists of only one connected component, that the Laplacian will have only one non-zero eigenvalue.

Let the number of clusters that the graphs should be separated into be k. Taking the k eigenvectors that correspond to the k eigenvalues closest to 0, and transforming the nodes of the graph into the k-dimensional space that they form, all the nodes that belong to one cluster will be situated close in that space. The new k-dimensional space is represented by matrix U, and each protein is represented by a single row in this matrix. We can cluster these $|V|$ points by using the standard k-means clustering algorithm. The number of clusters is determined empirically. The step by step algorithm is presented in Fig. 2.

Input: the interaction network $G = (V;E)$; number of clusters k;
Output: clusters $C_1, \ldots, C_k$;

Let $\mathbf{A}$ be the adjacency matrix defined by E;

Let D be a diagonal matrix whose diagonal element D_{ii} equals the degree of the node i of the graph;

1. Compute the unnormalized Laplacian $L=D-A$
2. Compute the first k generalized eigenvectors $u_1,\ldots, u_k$ of the generalized eigenproblem
   $Lu = \lambda Du$
3. Construct matrix $U \in R^{|V| \times k}$ containing the vectors $u_1,\ldots, u_k$ as columns
4. Let $y_i \in R^k$ be the vector corresponding to the i-th row of matrix U
5. Cluster the points $(y_i)_{i=1,\ldots,|v|}$ in R^k with the k-means algorithms into clusters
   $C_1, \ldots, C_k$

Fig. 2. The spectral clustering algorithm

2.2 Functional Annotation Using Clusters

After clustering the PIN we set up a strategy for annotating the query protein with the adequate functions according to the functions of the other proteins in the cluster where it belongs. The simplest and most intuitive approach would be that each function is ranked by its frequency of appearance as an annotation for the proteins in the cluster. This rank is calculated by the formula (3) and is then normalized in the range from 0 to 1.

$$f(j)_{j \in F} = \sum_{i \in K} z_{ij} \qquad (3)$$

where F is the set of functions present in the cluster K, and

$$z_{ij} = \begin{cases} 1, \text{if } i\text{-th protein from K is annotated with the } j\text{-th function from F} \\ 0, \text{otherwise} \end{cases} \qquad (4)$$

2.3 Protein-Protein Interaction Network Weighting

We performed network weighting based on the annotation correlation between nodes. With respect to a particular protein interaction (P1, P2), (a) two GO terms co-appear if one of the GO terms is assigned to P1 and the other is assigned to P2, (b) two GO terms are co-absent if none of the two GO-terms are assigned to P1 or P2, (c) two GO terms cross-appear if one of the GO terms is assigned to protein P1 and the other GO term is not assigned to P2 and vice-versa. Based on the frequencies of co-appearance, cross-appearance and co-absence we used their correlation as weights for the edges between nodes. The following correlation measures were employed:

- Jaccard $\qquad J=F_{11}/(F_{11}+F_{01}+ F_{10})$ $\qquad\qquad$ (5)
- Cosine $\qquad C=F_{11}/(F_{+1}+F_{1+})^{0,5}$ $\qquad\qquad$ (6)
- H-measure $\qquad H=1-(F_{10}*F_{01})/(F_{+1}*F_{+1})$ $\qquad\qquad$ (7)

- Support $Sup=F_{11}$ (8)
- Confidence $Conf=max(F_{11}/F_{1+},F_{11}/F_{+1})$ (9)

where F_{11} denotes the co-appearance frequency, F_{01} and F_{10} denote the cross-appearance frequencies, and F_{1+} and F_{+1} denote the sum of frequencies for terms appearing in one of the proteins connected with the edge we are weighting (i.e. $F_{1+}=F_{11}+F_{10}$).

3   Results and Discussion

High-throughput techniques are prone to detecting many false positive interactions, leading to a lot of noise and non-existing interactions in the databases. Furthermore, some of the databases are supplemented with interactions computationally derived with a method for protein interaction prediction, adding additional noise to the databases. Therefore, none of the available databases are perfectly reliable and the choice of a suitable database should be made very carefully.

For the needs of this paper the PIN data are compiled, pre-processed and purified from a number of established datasets, like: DIP, MIPS, MINT, BIND and BioGRID. The functional annotations of the proteins were taken from the SGD database [24]. It is important to note that the annotations are unified with Gene Ontology (GO) terminology [25].

The data for protein annotations are not used raw, but are preprocessed and purified. First, the trivial functional GO annotations, like 'unknown cellular compartment', 'unknown molecular function' and 'unknown biological process' are erased. Then, additional annotations are calculated for each protein by the policy of transitive closure derived from the GO. The extremely frequent functional labels (appearing as annotations to more than 300 proteins) are also excluded, because they are very general and do not carry significant information.

After all the preprocessing steps, the used dataset is believed to be highly reliable and consists of 2502 proteins from the interaction of the baker's yeast, has 12708 interactions between them and are annotated with a total of 888 functional labels. For the purposes of evaluating the proposed methods, the largest connected component of this dataset is used, which consists of 2146 proteins.

Each protein in the PIN is streamed through the prediction process one at a time as a query protein. The query protein is considered unannotated, that is we employ the leave-one out method. Each of the algorithms works in a fashion that ranks the "proximity" of the possible functions to the query protein. The ranks are scaled between 0 and 1 as explained in 2.2. The query protein is annotated with all functions that have rank above a previously determined threshold ω. For example, for $\omega = 0$, the query protein is assigned with all the function present in its cluster. We change the threshold with step 0.1 and compute numbers of true-positives (TP), true-negatives (TN), false-positives (FP) and false-negatives (FN). For a single query protein we consider the TP to be the number of correctly predicted functions, and for the whole PIN and a given value of ω the TP number is the total sum of all single protein TPs .

To compare performance between different algorithms we use standard measures as sensitivity and specificity (10).

$$sensitivity = \frac{TP}{TP + FN} \qquad specificity = \frac{TN}{TN + FP} \qquad (10)$$

We plot the values we compute for the sensitivity and specificity using a ROC curve (Receiver Operating Curve). The x-axe corresponds to the false positive rate, which is the number of false predictions that a wrong function is assigned to a single protein, scaled by the total number of functions that do not belong to that particular protein. This rate is calculated with (11).

$$fpr = \frac{FP}{FP + TN} = 1 - specificity \qquad (11)$$

The y-axe corresponds to the rate of true predictions that is the sensitivity. At last we use the AUC (Area Under the ROC Curve) measure as a numeric evaluator of the ROC curve. The AUC is a number that is equal to the area under the curve and its value should be above 0.5, which is the value that we get if the prediction process was random. The closer the value of AUC to 1, the better is the prediction method.

Table 1. Evaluation results of the unweighted spectral clustering method

number of eigenvalues	$\omega =$	0,1	0,3	0,5	0,7	0,9	AUC
50	sens.	0,6484	0,4436	0,3246	0,2082	0,1147	0,8644
50	fpr	0,0565	0,0159	0,0077	0,0036	0,0014	0,8644
100	sens.	0,6702	0,4713	0,3376	0,2430	0,1404	0,8590
100	fpr	0,0479	0,0142	0,0056	0,0028	0,0010	0,8590
150	sens.	0,6709	0,5053	0,3620	0,2688	0,1598	0,8531
150	fpr	0,0400	0,0128	0,0050	0,0026	0,0012	0,8531

Table 2. Evaluation results of the weighted spectral clustering method using Jaccard correlation measure

number of eigenvalues	$\omega =$	0,1	0,3	0,5	0,7	0,9	AUC
50	sens.	0,7262	0,4725	0,3394	0,2253	0,1332	0,9146
50	fpr	0,0645	0,0168	0,0071	0,0032	0,0013	0,9146
100	sens.	0,7482	0,5488	0,4017	0,2967	0,1788	0,9061
100	fpr	0,0469	0,0137	0,0052	0,0026	0,0011	0,9061
150	sens.	0,7482	0,5694	0,4238	0,3195	0,1915	0,8979
150	fpr	0,0384	0,0120	0,0047	0,0024	0,0010	0,8979

As mentioned before, the number of eigenvalues that were considered, i.e. the number of clusters that the PIN is segmented into is determined empirically.

Experiments were performed with segmenting the networks in 50, 100 and 150 clusters. The evaluation results for the unweighted spectral clustering are given in Table 1. Regarding the weighted spectral clustering, Table 2 shows the evaluation results using the Jaccard correlation measure only.

It is evident from the tables below, that when comparing the achieved AUC values, the both methods perform the best when the clusters are determined using only 50 eigenvalues. Furthermore, the weighted clustering scheme is evidently superior, improving the sensitivity by 7% while achieving the same (when 50 eigenvalues are considered) or even smaller (when 100 and 150 eigenvalues are considered) false positive rate. The AUC values also note significant improvement when using the weighting clustering scheme.

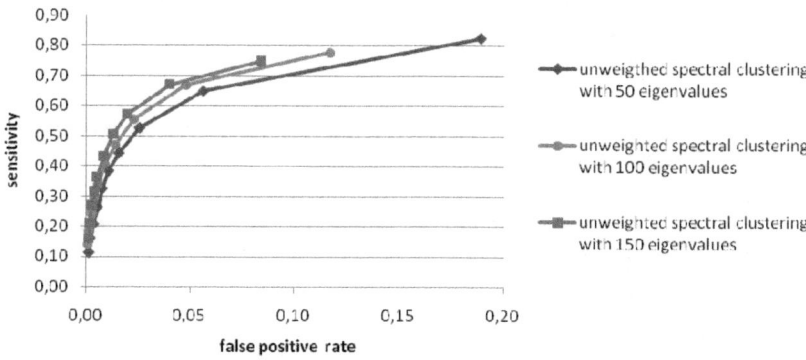

Fig. 3. ROC curves for the function prediction evaluation for the unweighted spectral clustering

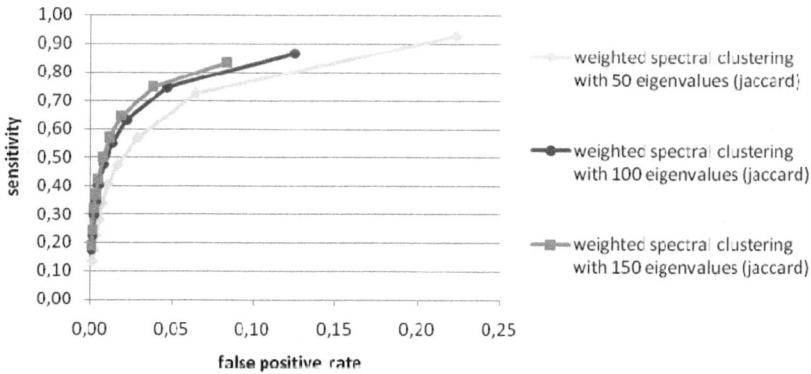

Fig. 4. ROC curves for the function prediction evaluation for the weighted spectral clustering using Jaccard correlation measure

Graphical visualization of the function prediction results is given in Fig. 3 for the unweighted and Fig.4 for the weighted clustering scheme. For $\omega = 0$, the methods reach the highest sensitivity of 82.35% and even 92.72% for the unweighted and weighted clustering scheme respectively. This is achieved for 50 eigenvalues considered for the both methods. However, the price is also very high false positive rate of nearly 20%. This is unacceptably high false positive rate, since the number of function one protein is not annotated with is many times bigger than the number of function it is annotated with. Exactly this is the reason why the both methods have the highest AUC values when 50 eigenvalues are considered. But, when comparing the results for $\omega > 0.1$, it is clear that the both methods perform the best when the PIN is separated into 150 clusters. In those cases the sensitivity is the highest, while the false positive rate is insignificantly small. According to this discussion, it would be always useful to inquire what the permissible trade-off limit between correctly and incorrectly detected protein functions is in order to determine what is the best method and best parameter to be used for protein function annotation.

The results using the other correlation measures: cosine, h-measure, support and confidence, are not given due to space limitations. For all the correlation measures only the AUC values are given and compared in Table 3. The highest AUC values are achieved when using the Jaccard correlation measure.

Table 3. Comparison of the AUC values of evaluation results of the weighted spectral clustering using different correlation measures

number of eigenvalues	Jaccard	Cosine	H-measure	Support	Confidence
50	0,9146	0.9116	0.9071	0.9011	0.9109
100	0,9061	0.9055	0.9025	0.8928	0.9042
150	0,8979	0.8955	0.8953	0.8823	0.8963

4   Conclusion

This paper exploits the ability of spectral clustering for detecting functional modules and predicting protein functions from PIN. The method was tested over one of the richest interactomes: the interactome of the baker's yeast. The method performs spectral clustering over the Laplacian of the adjacency matrix of the PIN. The algorithm was tested on unweighted and weighted PIN graph. In the weighting process we used annotation correlation between two neighbouring proteins in the PIN. Due to the fact that the PIN data contain a lot of false positive interactions, the dataset needed to be preprocessed and purified prior to the functional annotation. This paper also illustrates a general framework for the vast set of algorithms for protein function prediction which are based on clustering of the PIN. The proposed approach proves that utilizing clustering of the PIN has high potential in the task of protein function prediction. The results show that our algorithm achieves high sensitivity and small false positive rate on both weighted and unweighted PIN graphs and for both it has a high AUC value, with a clear advantage when using the weighting scheme.

References

1. von Mering, C., Krause, R., Sne, B., et al.: Comparative assessment of large-scale data sets of protein-protein interactions. Nature 417(6887), 399–403 (2002)

2. Hakes, L., Lovell, S.C., Oliver, S.G., et al.: Specificity in protein interactions and its relationship with sequence diversity and coevolution. PNAS 104(19), 7999–8004 (2007)
3. Harwell, L.H., Hopfield, J.J., Leibler, S., Murray, A.W.: From molecular to modular cell biology. Nature 402, c47–c52 (1999)
4. Brohée, S., van Helden, J.: Evaluation of clustering algorithms for protein-protein interaction networks. BMC Bioinformatics 7, 48 (2006)
5. Barabasi, A.L., Oltvai, Z.N.: Network biology: understanding the cell's functional organization. Nat. Rev. Genet. 5, 101–113 (2004)
6. Arnau, V., Mars, S., Marin, I.: Iterative cluster analysis of protein interaction data. Bioinformatics 21, 364–378 (2005)
7. Rives, A.W., Galitski, T.: Modular organization of cellular networks. PNAS 100, 1128–1133 (2003)
8. Friedel, C.C., Zimmer, R.: Inferring topology from clustering coefficients in protein-protein interaction networks. BMC Bioinformatics 7, 519 (2006)
9. Pereira-Leal, J.B., Enright, A.J., Ouzounis, C.A.: Detection of functional modules from protein interaction networks. Proteins 54, 49–57 (2004)
10. Dunn, R., Dudbridge, F., Sanderson, C.M.: The use of edge-betweenness clustering to investigate biological function in PINs. BMC Bioinformatics 6, 39 (2005)
11. Luo, F., Yang, Y., Chen, C.F., Chang, R., Zhou, J., et al.: Modular organization of protein interaction networks. Bioinformatics 23, 207–214 (2007)
12. Newman, M.E., Girvan, M.: Finding and evaluating community structure in networks. Phys. Rev. E. Stat. Nonlin. Soft. Matter Phys. 69, 026113 (2004)
13. Asur, S., Ucar, D., Parthasarathy, S.: An ensemble framework for clustering protein-protein interaction networks. Bioinformatics 23, i29–i40 (2007)
14. Bader, G.D., Hogue, C.W.: An automated method for finding molecular complexes in large protein interaction networks. BMC Bioinformatics 4, 2 (2003)
15. King, A.D., Przulj, N., Jurisica, I.: Protein complex prediction via cost-based clustering. Bioinformatics 20, 3013–3020 (2004)
16. Spirin, V., Mirny, L.A.: Protein complexes and functional modules in molecular networks. PNAS 100(21) (2003)
17. Gagneur, J., Krause, R., Bouwmeester, T., Casari, G.: Modular decomposition of protein-protein interaction networks. Genome. Biol. 5, R57 (2004)
18. Morrison, J.L., Breitling, R., Higham, D.J., Gilbert, D.R.: A lock-and-key model for protein-protein interactions. Bioinformatics 22, 2012–2019 (2006)
19. Andreopoulos, B., An, A., Wang, X., Faloutsos, M., Schroeder, M.: Clustering by common friends finds locally significant proteins mediating modules. Bioinformatics 23, 1124–1131 (2007)
20. Royer, L., Reimann, M., Andreopoulos, B., Schroeder, M.: Unraveling protein networks with power graph analysis. PLoS Comput. Biol. 4, e1000108 (2008)
21. Belkin, M., Niyogi, P.: Laplacian Eigenmaps for Dimensionality Reduction and Data Representation. Neural Computation 15, 1373–1396 (2003)
22. Chen, J., Yuan, B.: Detecting Functional Modules in the Yeast Protein-Protein Interaction Network. Bioinformatics 18(22), 2283–2290 (2006)
23. Lancichinetti, A., Fortunato, S., Radicchi, F.: Benchmark Graphs for testing Community Detection Algorithms. Physical Review E78, 046110 (2008)
24. Dwight, S., Harris, M., Dolinski, K., Ball, C., Unkley, G.B., Christie, K., Fisk, D., Issel-Tarver, L., Schroeder, M., Sherlock, G., Sethuraman, A., Weng, S., Botstein, D., Cherry, J.M.: Saccharomyces Genome Database (SGD) provides secondary gene annotation using Gene Ontology (GO). Nucleic Acids Research 30(1) (2002)
25. The gene ontology consortium: Gene ontology: Tool for the unification of biology. Nature Genetics 25(1), 25–29 (2000)
26. Fortunato, S.: Community Detection in Graphs. Physics Reports 486, 75–174 (2010)

Reducing the Subjectivity of Gene Expression Data Clustering Based on Spatial Contiguity Analysis

Hui Yi[1], Xiaofeng Song[1,*], Bin Jiang[1], and Yufang Liu[1,2]

[1] Department of Biomedical Engineering, Nanjing University of Aeronautics and Astronautics, Nanjing, China
xfsong@nuaa.edu.cn
[2] Guodian Environment Protection Research Institute, Nanjing, China

Abstract. Clustering, which has been widely used as a forecasting tool for gene expression data, remains problematic at a very deep level: different initial points of clustering lead to different processes of convergence. However, the setting of initial points is mainly dependent on the judgments of experimenters. This subjectivity brings problems, including local minima and an extra computing consumption when bad initial points are selected. Hence, spatial contiguity analysis has been implemented to reduce the subjectivity of clustering. Data points near the cluster centroids are selected as initial points in this paper. This accelerates the process of convergence, and avoids the local minima. The proposed approach has been validated on benchmark datasets, and satisfactory results have been obtained.

Keywords: Clustering, Gene expression, Subjectivity, Initial points setting.

1 Introduction

Clustering is a widely used tool in genomic studies [1-5]. By clustering gene expression data with existed samples, the potential co-regulation or discriminate pathologies could be extracted. Although it has been studied for years and many remarkable achievements have been obtained [6-9], gene expression data clustering still remains problematic at a very deep level.

As Jain et al [7] has pointed out, clustering is a subjective process; The same set of data items often needs to be partitioned differently for different applications. Moreover, even for the same application, different selections of initial points lead to different classifications of gene expression data. The clustering results are highly dependent on the judgment of experimenters, and this subjectivity makes the process of clustering difficult. A misleading biological conclusion is likely to be obtained if a bad selection has been made arbitrarily.

From this perspective, subjectivity reduction is one of the most urgent problems that gene expression data clustering analysis is facing. Generally, it is important to run the clustering algorithms several times with different random seeds [8]. However, this random seeds approach requires much more computation and offers no theoretical

---

[*] Corresponding author.

T.-h. Kim et al. (Eds.): DTA/BSBT 2011, CCIS 258, pp. 118–124, 2011.
© Springer-Verlag Berlin Heidelberg 2011

guide for initial point selections. Moreover, it is not guaranteed to avoid the local minima unless the clustering is repeated for enough times. Hence reasonable criterions are required for initial point selection.

In this paper, we proposed a spatial contiguity based approach to select the initial points. This approach takes the idea that only the samples that from different clusters treated as the initial points could the issue of local minima be avoided, and the extra computing consumption eliminated.

2   Problem Formulation

Clustering can be regarded as an iterating process that initial points converge to the centroids of corresponding clusters [10, 11]. Hence the convergence will be quick, and is hard to fall into the local minima if the data items near by the centroid positions are selected as initial points. Oppositely, it requires more computation to achieve the convergence, or even only a wrong convergence could be obtained. The following example is designed as an illustration. Given a group of 2 dimension samples, which is shown in Tab.1, it contains 17 samples within 3 types. The distribution is shown in Fig.1.

Table 1. Dataset for illustrating Example

Sample	Value	Sample	Value	Sample	Value	Sample	Value	Sample	Value
X1	(2,2)	x2	(2,3)	x3	(4,2)	x4	(4,3)	x5	(3,5)
X6	(4,5)	x7	(4.5,6)	x8	(5.5,5)	x9	(6,7)	x10	(7,6)
X11	(7.5,7)	x12	(8.5,7)	x13	(6,4)	x14	(6,3)	x15	(7,4)
X16	(7.5,3)	x17	(8,3.5)						

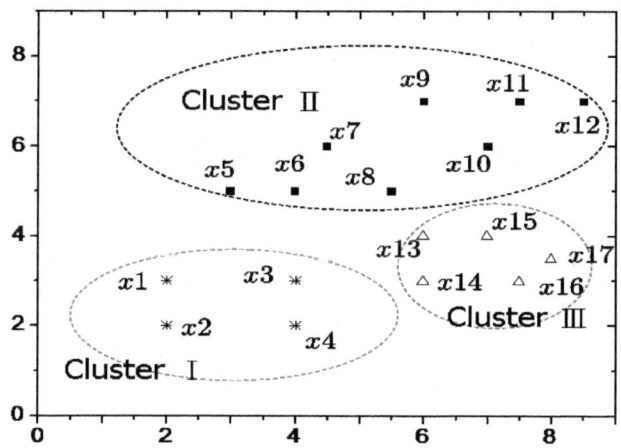

Fig. 1. Spatial Distribution of The 2-dimension Data Items

Table 2. The Clustering Performances with Different Initial Points

Selection of Initial Points		(x_1, x_2, x_3)	(x_8, x_9, x_{10})	(x_7, x_8, x_9)	(x_1, x_{12}, x_{16})
Iteration times		4	5	4	2
Total distances within clusters		39.8375	39.8375	32.2	32.2
Results of clustering	Cluster I	x_1, x_2, x_3, x_4	x_1, x_2, x_3, x_4	x_1, x_2, x_3, x_ι	x_1, x_2, x_3, x_4
	Cluster II	x_5, x_6, x_7, x_8	x_5, x_6, x_7, x_8	$x_5, x_6, x_7, x_8,$ x_9, x_{10}, x_{11}, x_1	$x_5, x_6, x_7, x_8,$ $x_9, x_{10}, x_{11}, x_{\cdot}$
	Cluster III	$x_9, x_{10}, x_{11}, x_{12}$ $x_{13}, x_{14}, x_{15}, x_1$ x_{17}	$x_9, x_{10}, x_{11}, x_{12}$ $x_{13}, x_{14}, x_{15}, x_1$ x_{17}	$x_{13}, x_{14}, x_{15}, x_1$ x_{17}	$x_{13}, x_{14}, x_{15}, x_1$ x_{17}

As shown in Tab. 2, the selections of (x1, x2, x3) and (x8, x9, x10) as initial points have fallen into local minima. The selection of (x7, x8, x9) has a correct convergence, but the computing consumption is too large. While using (x1, x12, x16) as the initial points, the clustering converges correctly and rapidly. Evidently, the selection of initial points has direct relationships with the converging correctness and speed.

Traditionally, experimenters can choose a best selection of initial points by a 'random seed' method [8]. This method prepares different groups of initial points, and calculates the squares of clusters for each group. Then it takes the group with the least squares as the best selection. The effectiveness of this approach depends on the number of groups. A mass of computations are required to guarantee the clustering avoiding the local minima. And it doesn't offer any theoretical guide for initial point selection.

3   Initial Points Selection Based on Spatial Contiguity Analysis

The best setting of initial points should be the centriods. However, centriods could not been detected until the clustering is completed. Hence, sub-optimal ones which are regarded near the centriods by spatial contiguity analysis are implemented as initial points in our approach. It could be deduced from Tab.2 that these points have two characteristics: (1) Initial points should be far from each other. For each cluster, one initial point should be selected; (2) It is better for the initial points being in the 'corner', for example, (x_1, x_{12}, x_{16}) are selected for Fig.1.

For 2-Dimension dataset, the initial points can be selected directly from figures. But gene expression datasets are usually high dimensional, and it is impossible for experimenters to find those 'corner' points directly. Hence the Principal Components

Analysis (PCA) which could reduce the dimension of given samples is introduced for spatial contiguity analysis.

Given the gene expression data $X \in \mathfrak{R}^{m \times n}$, firstly a new matrix Y is generated :

$$Y \equiv \frac{1}{\sqrt{n-1}} X^T \tag{1}$$

Then eigenvectors for $Y^T Y$ are calculated by (2)

$$\left| \lambda I - Y^T Y \right| = 0 \tag{2}$$

And (3) could also be obtained:

$$\begin{pmatrix} \lambda_i - 1 & -r_{12} & \cdots & -r_{1p} \\ -\lambda_{21} & \lambda_{i-1} & \cdots & -r_{2p} \\ & & \cdots & \\ -r_{n1} & -r_{n2} & \cdots & \lambda_{i-1} \end{pmatrix} \begin{pmatrix} \alpha_{1i} \\ \alpha_{2i} \\ \vdots \\ \alpha_{ni} \end{pmatrix} = \begin{pmatrix} 0 \\ 0 \\ \vdots \\ 0 \end{pmatrix} \tag{3}$$

where $\lambda_1 \geq \lambda_2 \geq \cdots \geq \lambda_n$, and $F_i = \alpha_{1i} Y_1 + \alpha_{2i} Y_2 + \cdots + \alpha_{ni} Y_n$ is the i-th principle component for gene expression data X. We use (4) to judge the importance of the first and second principle components.

$$R = \frac{\lambda_1 + \lambda_2}{\sum_{i=1}^{n} \lambda_i} \tag{4}$$

When the value of R is obtained, the selection of initial points could be made according to the following rules.

Case 1: $R < 0.3$

It means that there are no predominant directions in the spatial distribution of data. The spatial distribution of data is like a hyper-ball. Taking a 2-Dimension dataset for instance, the spatial distribution of data is like a circle. If we want to divide the data into k clusters, just make a maximum k-polygon inscribed in the circle, and the points corresponding to all apexes could be chosen as initial points of clustering.

Case 2: $R \geq 0.3$, Given :

$$\log_2 k \leq l < \log_2 k + 1, \ l \in N \tag{5}$$

where k is the number of clusters, and l is the number of principle components that should be used. We search for the samples with the biggest and smallest values for each principal component. Then 2^l samples $S = (q_1, q_2, \ldots, q_{2^l})$ could be obtained. The selecting process for initial points is shown in Tab.3. Fig.2 shows the initial points selection for the dataset in Tab.1, which is consistent with the results in Tab.2.

Table 3. The selection of initial points when $R \geq 0.3$

if k is odd

  if $\displaystyle\sum_{i=1}^{2^{l-1}} |q_{(2^l-1)} - q_i| - \sum_{i=1}^{2^{l-1}} |q_{2^l} - q_i| < 0$

    $InitialPoints = (q_1, q_2, \cdots, q_{2^{(l-1)}}, q_{2^l})$

  else

    $InitialPoints = (q_1, q_2, \cdots, q_{2^{(l-1)}}, q_{(2^l-1)})$

  end

 else

    $InitialPoints = S$

end

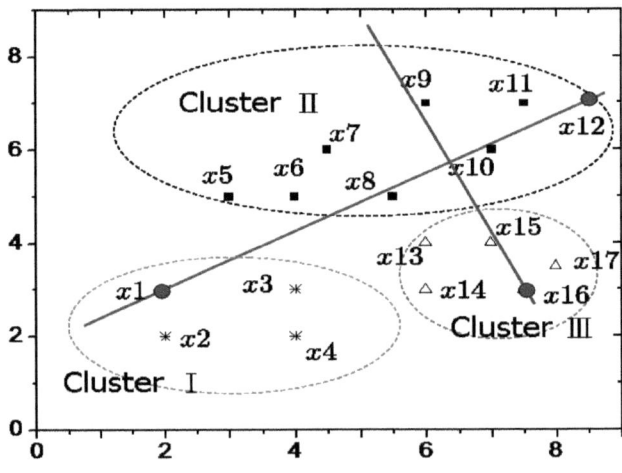

Fig. 2. Initial points Selection for illustrating example using Spatial Contiguity Analysis

4   Validations

Three datasets with different number of clusters, namely, Rats CNS dataset (k=2), B-cell Lymphoma dataset (k=5) and Yeast Saccharomyces Cerevisiae dataset (k=7), have been employed for validating the proposed approach. In the experiments, different initial points selected by both random seeds and our approach are implemented in the k-means clustering. And for each clustering, the iteration times and corresponding total distances of within cluster elements are recorded.

In Fig.3(a), all selections have converges at the same classification results. However, the approach based on spatial contiguity analysis takes just two iterations to achieve the convergence, and it converges much quicker than random selections. In Fig.3(b), a more complicated case has been investigated, where there has 5 clusters ($k = 5$), and the proposed approach has also made the quickest convergence. And in Fig.3(c), where there has 7 clusters, all random seeds selections have failed to achieve the smallest value of total point-to-centroid distances. This means they have fallen into local minima. Meanwhile, the spatial contiguity analysis approach has overcame the local minima, whilst yields a good converging speed.

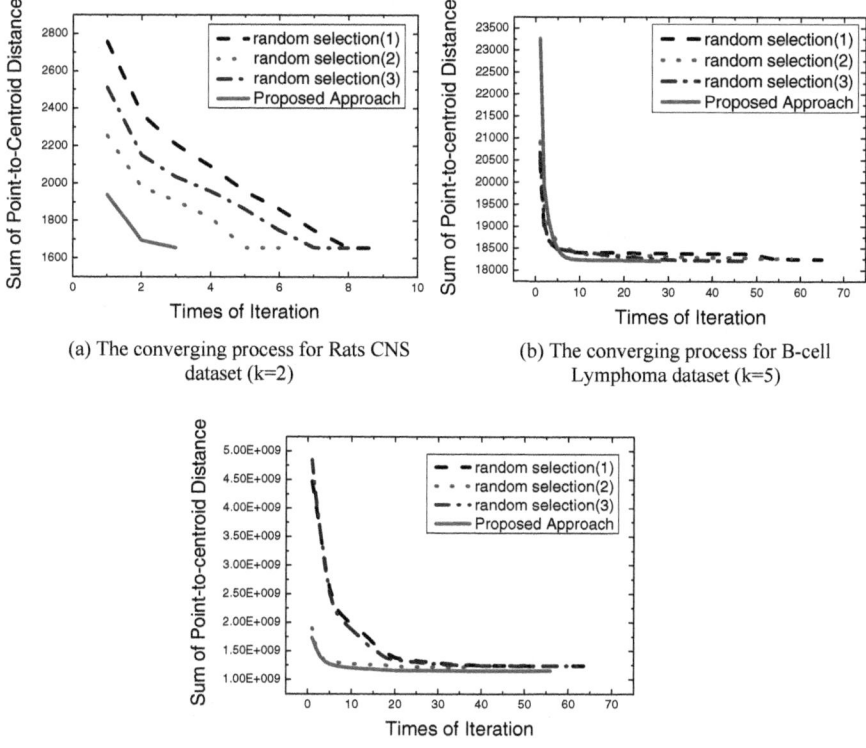

(a) The converging process for Rats CNS dataset (k=2)

(b) The converging process for B-cell Lymphoma dataset (k=5)

(c) The converging process for Yeast Saccharomyces Cerevisiae dataset (k=7)

Fig. 3. Comparative Experiments of Clustering With Different Initial Points Selection

In all different cases, the proposed approach is capable to select a reasonable set of initial points, and it shows the superiority to other selections. This demonstrates the effectiveness of the spatial contiguity based selection approach.

5   Conclusions

Clustering gene expression data is an important means for biological information extracting. But the process of clustering is subjective, and this may lead to misleading biological conclusions. It is an urgent requirement that the subjectivity of gene expression data clustering process should be reduced. Hence, we proposed a spatial contiguity based approach to select the initial points. This approach attempts to find samples that near by the centriods as initial points, which helps to accelerate the converging speed and avoid the problem of local minima. The proposed approach has been applied to three gene expression datasets with different number of clusters, and satisfactory results have been obtained.

Acknowledgements. The study was supported by grants from National Natural Science Foundation of China (No. 61171191) and Natural Science Foundation of Jiangsu Province in China (BK2010500).

References

[1] Yeung, K.Y., Haynor, D.R., Ruzzo, W.L.: Validating clustering for gene expression data. Bioinformatics 17(4), 309–318 (2001)
[2] Cho, R.J., Campbell, M.J., Winzeler, E.A., et al.: A genome-wide transcriptional analysis of the mitotic cell cycle. Molecular Cell 2, 65–73 (1998)
[3] Bolshakova, N., Azuaje, F.: Cluster validation techniques for genome expression data. Signal Processing 83, 825–833 (2003)
[4] Handl, J., Knowles, J., Kell, D.B.: Computation cluster validation in post-genomic data analysis. Bioinformatics 21(15), 3201–3212 (2005)
[5] Dougherty, E.R., Barrera, J., Brun, M., et al.: Inference from clustering with application to gene-expression microarray. J. Comput. Biol. 9(1), 105–126 (2002)
[6] Xu, R., Wunsch ll, D.: Survey of Clustering Algorithms. IEEE Trans. on Neural Networks 16(3), 645–678 (2003)
[7] Jain, A.K., Murty, M.N., Flynn, P.J.: Data clustering: a review. ACM Comput. Surveys 31(3), 264–323 (1999)
[8] D'haeseleer, P.: How does gene expression clustering work? Nature Biotechnology 23(12), 1499–1501 (2005)
[9] Dougherty, E.R., Brun, M.: A probabilistic theory of clustering. Pattern Recognition 37, 917–925 (2004)
[10] Bezdek, J.: Cluster validity with fuzzy sets. J. Cybernt. 3(3), 58–72 (1974)
[11] Bezdek, J., Hathaway, R., et al.: Convergence theory for fuzzy c-means: counterexamples and repairs. IEEE Transactions on Systems, Man and Cybernetics 17(15), 873–877 (1987)

A Novel Non-contact Infection Screening System Based on Self-Organizing Map with K-means Clustering

Guanghao Sun[1], Shigeto Abe[2], Osamu Takei[3], Yukiya Hakozaki[4],
and Takemi Matsui[1]

[1] Department of Management Systems Engineering, Tokyo Metropolitan University,
Asahigaoka 6-6, Hino, Tokyo, Japan
{sun-guanghao,tmatsui}@sd.tmu.ac.jp
http://www.sd.tmu.ac.jp/matsui-lab/takemi_matsui.html
[2] TAKASAKA Clinic, Kanesaka 172-21, Uchigomiya, Iwaki, Fukushima, Japan
[3] LIFETECH CO., Ltd, Miyadera 4074, Iruma, Saitama, Japan
[4] Department of Internal Medicine, Japan Self-Defense Forces Central Hospital,
1-2-24 Ikejiri, Setagaya, Tokyo, Japan

Abstract. This paper aims to evaluate the efficacy of our non-contact infection screening system which uses Kohonen's self-organizing map (SOM) with K-means clustering algorithm. In this study, the linear discriminant analysis (LDA) used in our previous system was replaced by SOM with K-means clustering algorithm to increase accuracy. The system simultaneously measures heart rate, respiratory rate, and facial skin temperature. The evaluation was done using the same data which we used in our previous study. The data was based on the test on 57 influenza patients and 35 normal control subjects at Japan Self-defense Forces Central Hospital. The system showed higher sensitivity of 98% and negative predictive value (NPV) of 96% compared to our previous system (sensitivity of 89%, NPV of 83%). The system can be used as a public health measure at points of entry where high sensitivity is most required in order to prevent the spread of the pandemic.

Keywords: Screening, infection, self-organizing map, K-means, thermography, heart rate, respiratory rate, microwave radar.

1   Introduction

Following the outbreak of severe acute respiratory syndrome (SARS) worldwide, public health measures such as screening of infected passengers at points of entry have been implemented in order to prevent the pandemic [1, 2]. Since fever is one of the major symptoms of SARS and influenza, infrared thermography is adopted for febrile passenger screenings by monitoring their skin surface temperatures in many countries [3]. However, the skin surface temperature measured by thermography is affected by many factors, such as antifebrile intake, alcohol drinking and ambient temperatures [4]. A previous study indicated that fever alone may be insufficient as a parameter to detect infected individuals at border quarantines [5].

T.-h. Kim et al. (Eds.): DTA/BSBT 2011, CCIS 258, pp. 125–132, 2011.
© Springer-Verlag Berlin Heidelberg 2011

In order to achieve more accurate screening, we have developed a screening system, which monitors infection-induced alternation of heart rates and respiratory rates as well as skin surface temperature in our previous study [6, 7]. These parameters increase by outbreak of infectious diseases excluding cholera. By adding these new parameters, the system showed higher screening accuracy than that using thermography alone. The system is composed of thermography to measure facial skin temperature, a laser Doppler blood-flow meter to measure heart rates and 10-GHz microwave radar to measure respiratory rates, and then distinguishes influenza group from normal group via linear discriminant analysis (LDA) using derived variables within several tens of seconds.

LDA is a linear supervised dimensionality reduction method. LDA try to find the optimal discriminant function by maximizing the between-class scatter and minimizing the within-class scatter [8]. However, LDA tends to give unwanted clustering results if the samples in a class are multimodal data (heart rates, respiratory rates and facial skin temperature) [9]. The infection screening system based on LDA with the sensitivity is less than 90%, this can be attributed to the limitation of LDA. Sensitivity is the most important parameter in quarantine screening from a viewpoint of infection prevention. For this reason, there is a need for an unsupervised discriminant function to improve the sensitivity of infection screening.

In this paper, we proposed a discriminant method using Kohonen's self-organizing map combined with K-means clustering algorithm. SOM can classify the derived variables based on the unsupervised classification and self-organizing learning [10], the proposed method can optimize the discriminant function by itself without using supervisor in real time. We adopted SOM as a first step clustering method using feature vectors, such as heart rates, respiratory rates and facial skin temperature. After training the SOM, several clusters are color-coded. A method for differentiating influenza group and normal group is a two-class problem. K-means algorithm was employed in order to reduce the number of clusters to two.

In order to verify the usefulness of the proposed method, SOM with K-means based infection screening system was tested using the previously obtained data for our previous paper [7]. The data was based on the test which had been performed on 57 medicated influenza patients and 35 normal control subjects at Japan Self-defense Forces Central Hospital. Moreover, we also compared the performance of the proposed method with LDA.

2   Methodology

2.1   Non-contact Screening System Hardware Setup

The technical details of the non-contact infection screening system have been reported in our previous paper [6]. The screening system simultaneously measures heart rates, respiratory rates and facial skin temperature. The system consists of a non-contact laser Doppler blood-flow meter, a 10-GHz respiration radar and an infrared thermography. The schematic diagram of the non-contact screening system is shown in Fig. 1. The outputs of the laser Doppler blood-flow meter, 10-GHz respiration radar, and infrared thermography were transferred to a personal laptop computer, then analyzed and displayed in real time. The screening program incorporates band pass filters for monitoring respiratory and heart rates in order to eliminate background noise. The heart rate is obtained by fast Fourier transform (FFT), and the respiratory rate is determined by the respiration curve measured by the respiration radar. The program is written in LabVIEW (National Instruments, USA) and SOM toolbox for MATLAB (Mathworks, USA).

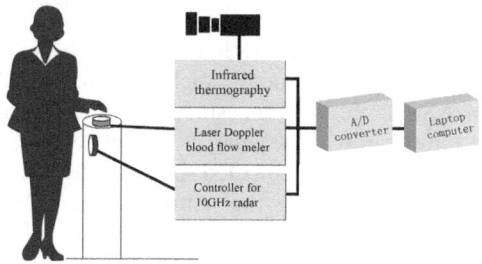

Fig. 1. The diagrammatic illustration of the non-contact infection screening system

2.2   Self-Organizing Map with K-means Two-Layer Clustering Method

A self-organizing map was developed by Kohonen in 1980 [10], which is a model using unsupervised learning neural network. SOM's training algorithm is based on competitive learning, which calculates the distance between an input vector $x(t)$ and all the weight vectors $m_i(t)$. The weight vector is closest to the input vector $x(t)$, which is called the Best-Matching Unit (BMU). The winner neuron can be determined as follow:

$$\|x(t) - m_c(t)\| = \min\{\|x(t) - m_i(t)\|\}. \tag{1}$$

where c is winner neuron, $\|\bullet\|$ is the distance measure. After finding the BMU, the weight vectors are updated to move the BMU closer to the input vector. The update rule for each neuron i is given by:

$$m_i(t+1) = m_i(t) + \alpha(t)[x(t) - m_i(t)]. \tag{2}$$

where $\alpha(t)$ is the learning rate, $x(t)$ is the input vector. After training the map, several clusters are visualized using the unified distance matrix (U-matrix). The U-matrix shows distances between neighboring map units using color levels [11]. High values on the U-matrix mean large distance between neighboring map units, and thus indicate cluster borders. By using the U-matrix, we can find out the distribution of clusters on the map. In order to separate the clusters optimally by using the U-matrix, we adopted K-means algorithm.

K-means algorithm calculates the closest distance between the centroid of current clusters and an input vector, and then assigns input vectors to the cluster. Then the new cluster's centroid is re-calculated, until the location of centroid no more changed

[12]. The distance between input vector and cluster centroid is calculated by the following equation:

$$D = \sum_{i=1}^{C} \|x - c_i\|^2 \tag{3}$$

where D is a distance between a data point x, C is the number of clusters, c_i is the centroid of the cluster. The proposed SOM with K-means two-layer clustering method in infection screening system consists of three steps, as stated below.

Step1: Dataset construction and normalization
First, the data set consisting of samples such as normal subjects and medicated patients is read from ASCII file. The measured variables are heart rates, respiratory rates and facial skin temperature. The label is associated with each sample, normal control subjects as "NOR" and medicated patients as "INF". The scale of the variables is important in determining the nature of the SOM. For this reason, we normalize the all variables with the logarithmic scaling.

Step2: SOM training and visualization
The normalized data set is analyzed using the SOM, linear initialization and batch training algorithms are used. After training the map, several clusters are visualized on U-matrix. The U-matrix visualized the distances between neighboring map units using color levels.

Step3: K-means over the trained SOM
K-means clustering algorithm (C = 2) is applied over the trained SOM in order to bring down several clusters to two clusters and improve the quality of map visualization.

2.3   The SOM with K-means Clustering Method Test on the Clinical Data

We tested the SOM with K-means clustering method using the formerly obtained data for our previous paper [7]. The data was based on the test which had been performed on 57 medicated influenza patients (35.7 °C ≤ body temperature ≤ 38.3 °C, 19 to 40 years) and 35 normal control subjects (35.5 °C ≤ body temperature ≤ 36.9 °C, 21 to 35 years) at Japan Self-defense Forces Central Hospital. The study was reviewed and approved by the Ethics Committee of Japan Self-defense Forces Central Hospital. We made a percomparison between proposed SOM with K-means method and that of LDA.

3   Results

3.1   SOM with K-means Clustering Method Evaluation on the Clinical Data

The clinical data set was created for three variables in an ASCII file as shown in Table 1. The ASCII file is consisted of 35 normal control subjects (NOR1, NOR2...)

and 57 influenza patients (INF1, INF2…). Logarithmic processing was conducted for these three parameters, that is, facial skin temperature, respiratory rate and heart rate. The logarithmic processing results are normalized.

Table 1. Part of the clinical data set, normal subject is labeled as 'NOR' and influenza subjects is labeled as 'INF'

	Label	1st variable ↓ facial temperature [°C]	2nd variable ↓ respiratory rates[bpm]	3rd variable ↓ heart rates[bpm]
1st sample →	NOR1	32.3	13	64
2nd sample →	NOR2	32.4	15	70
3rd sample →	INF1	33.5	17	76
etc.	INF2	32.3	15	75

We conducted SOM training to cluster normal and infected subjects using MATLAB. The clustering results are shown in Fig. 2. The map unit of U-matrix determined by SOM are visualized according to the distances between neighboring map units using color levels (Fig. 2, Left). Red and blue hexagonal shape represent long and short distances respectively. The map units corresponding to normal control subjects tend to be distributed in the upper part and that corresponding to influenza patients have a tendency to be found in the lower part.

In order to optimally classify U-matrix into two clusters (normal control group and influenza group), K-means clustering algorithm was applied over trained SOM (Fig. 2, Right). A total of 56 out of 57 influenza patients are contained in the light blue cluster, 27 out of 35 normal control subjects are found in the dark blue cluster.

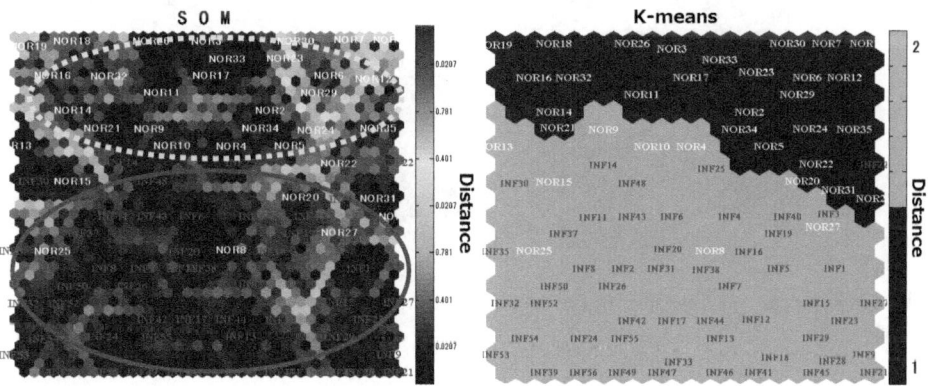

Fig. 2. Left. The color coding of SOM results (U-matrix), a fair portion of influenza patients are bounded by the red solid ellipse and a fair portion of normal subjects are bounded by the yellow dotted ellipse. Right. K-means clustering algorithm is applied over the trained SOM, the map is optimally divided into two clusters. A total of 56 out of 57 influenza patients are contained in the light blue cluster, 27 out of 35 normal control subjects are contained in the dark blue cluster.

3.2   The Comparison of SOM with K-means and LDA Diagnostic Test

Screening accuracy of the proposed method is compared with that of LDA method presented in our previous paper [7] (Table 2). For medicated influenza patients, the non-contact screening system using SOM with K-means clustering algorithm showed the sensitivity of 98% and negative predictive value (NPV) of 96%, the sensitivity and NPV corresponding to LDA are 89% and 83%, respectively. SOM with K-means improved the sensitivity and NPV drastically. This improvement is remarkable if we consider the fact that about half of influenza patients have no fever with antifebrile use. Sensitivity and NPV are the most important parameters in quarantine screening from a viewpoint of infection prevention. Whereas, the specificity and positive predictive value (PPV) of SOM with K-means were lower than those of LDA.

Table 2. Comparison of the diagnostic test results of the SOM with K-means method and LDA

Method	Sensitivity	Specificity	PPV	NPV
SOM with K-means	98%	77%	87%	96%
LDA	89%	85%	91%	83%

A sample of screening result using SOM with K-means is shown in Figure 3. The screening result "PASS", thermograph, respiratory curve, pulse curve are displayed at a laptop graphic terminal.

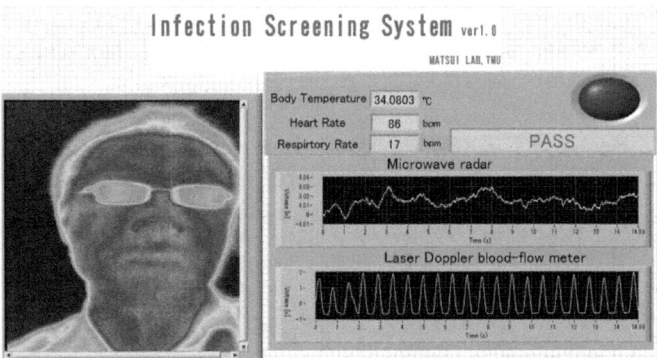

Fig. 3. Graphical user interface of the non-contact infection screening system using SOM with K-means

4   Discussions

A thermography based screening method has been adopted in detecting influenza and other infectious diseases at places of mass gathering, such as, international airports, hospital outpatient units and public health centers [13]. Some recent studies revealed that a screening method based on thermography alone does not guarantee sufficient sensitivity [5]. In order to achieve higher accuracy in screening infected individuals, we have developed a novel screening system which monitors respiratory rate and heart rate as well as well surface skin temperature. This system improved screening sensitivity compared to thermography-dependent screening system.

In our previous study, we employed the linear discriminant analysis (LDA) to distinguish the influenza group from measured parameters, i.e., heart rates, respiratory rates and surface skin temperature. LDA enables classification by maximizing the between-class distance while minimizing the within-class distance [9]. The sensitivity of the system using LDA did not exceed 90 %. The sensitivity is the most important parameter in infection screening. Low sensitivity means more entries of infected passengers into a country, which may lead to the pandemic spread of the disease. LDA also has a limitation on the ground that LDA parameters are expected to be independent and normally distributed.

The SOM with K-means clustering algorithm achieved high screening accuracy, i.e., the sensitivity of 98%, the sensitivity of great account is higher than that of LDA, whereas the specificity was lower than that of LDA. The sensitivity and specificity have the trade-off relationship in a screening situation [14]. SOM with K-mean clustering algorithm distinguishes the influenza group from normal group more accurately than LDA. SOM is an unsupervised neural network model, SOM with K-means clustering algorithm optimizes the discriminant function by itself without using supervisors in real time.

As a limitation of SOM with K-means algorithm, it provides the lower specificity than that of LDA. Lower specificity means higher false positive, higher false positive induces higher quarantine officers' workloads.

Other classification algorithms such as support vector machines should be tested in order to conduct more accurate infection screening.

References

1. World Health Organization: Communicable Disease Surveillance and Response: Severe Acute Respiratory Syndrome (SARS): Status of the outbreak and lessons for the immediate future, Geneva (2003)
2. St John, R.K., King, A., de Jong, D., Bodie-Collins, M., Squires, S.G., Tam, T.W.: Border screening for SARS. Emerg. Infect. Dis. 11, 6–10 (2005)
3. Chan, L.S., Cheung, G.T., Lauder, I.J., Kumana, C.R., Lauder, I.J.: Screening for fever by remote-sensing infrared thermographic camera. J. Travel Med. 11, 273–279 (2004)
4. Liu, C.C., Chang, R.E., Chang, W.C.: Limitations of forehead infrared body temperature detection for fever screening for severe acute respiratory syndrome. Infect. Control Hosp. Epidemiol. 25, 1109–1111 (2004)
5. Nishiura, H., Kamiya, K.: Fever screening during the influenza (H1N1-2009) pandemic at Narita International Airport, Japan. BMC Infectious Diseases 11, 111 (2011), doi:10.1186/1471-2334-11-111
6. Matsui, T., Suzuki, S., Ujikawa, K., Usui, T., Gotoh, S., Sugamata, M.: The development of a non-contact screening system for rapid medical inspection at a quarantine depot using a laser Doppler blood-flow meter, microwave radar and infrared thermography. J. Med. Eng. Technol. 33(6), 481–487 (2009)
7. Matsui, T., Hakozaki, Y., Suzuki, S., Usui, T., Kato, T., Hasegawa, K., Sugiyama, Y., Sugamata, M., Abe, S.: A novel screening method for influenza patients using a newly developed non-contact screening system. J. Infect. 60, 271–277 (2010)

8. Fukunaga, K.: Introduction to statistical pattern recognition. Academic Press, Tokyo (1990)
9. Sugiyama, M.: Dimensionality Reduction of Multimodal Labeled Data by Local Fisher Discriminant Analysis. The Journal of Machine Learning Research 8, 1027–1061 (2007)
10. Kohonen, T.: The self-organizing map. Proceedings of the IEEE 78, 1464–1480 (1990)
11. Vesanto, J., Himberg, J., Alhoniemi, E., Parhankangas, J.: Self-organizing map in Matlab: the SOM Toolbox. In: Proceedings of the Matlab DSP Conference, pp. 16–17 (1999)
12. MacQueen, J.: Some methods for classification and analysis of multivariate observations. In: Proceedings of 5th Berkeley Symposium on Mathematical Statistics and Probability, pp. 281–297 (1967)
13. Bitar, D., Goubar, A., Desenclos, J.C.: International travels and fever screening during epidemics: a literature review on the effectiveness and potential use of non-contact infrared thermometers. Eurosurveillance 12, 1–5 (2009)
14. Lalkhen, A.G., McCluskey, A.: Clinical tests: sensitivity and specificity. Continuing Education in Anaesthesia, Critical Care & Pain 2008 8, 221–223 (2008)

Integrating Inductive Knowledge into the Inference System of Biomedical Informatics

Kittisak Kerdprasop and Nittaya Kerdprasop

Data Engineering Research Unit, School of Computer Engineering,
Suranaree University of Technology, 111 University Avenue,
Nakhon Ratchasima 30000, Thailand
{kerdpras,nittaya}@sut.ac.th

Abstract. This paper presents a new methodology for the design and implementation of the next generation rule-based expert system in a medical domain. In addition to the set of predefined rules, the system includes rules that are automatically induced from the database instances. We design the inductive expert system such that the inductive process has been done through the tree-based knowledge discovery technique. Probabilistic decision rules are then transformed from the induced decision tree. The induced, as well as predefined, rules together form a knowledge base for the inductive expert system. Another feature of our system is the inference engine that can be created automatically. The system is intended to support decision making in biomedical informatics. The general design of our system is, however, appropriate for other domains as well.

Keywords: Automatic knowledge acquisition, Knowledge mining, Inductive expert system, Medical decision support system.

1 Introduction

Computers have been applied to medicine and health care since the 1950s as significant tools in medical diagnosis and therapy [22]. The success of medical expert systems such as MYCIN [24] and INTERNIST-1 [14] has attracted considerable attention from cross-discipline researchers including medical experts, computer scientists, engineers, decision analysts, and mathematicians. Later development of applications such s electronic health records, hospital information system, medical decision support systems, and many others have contributed to the emergence of medical informatics as a new academic discipline. The term *medical informatics* was originally coined in Europe to address the focus on application of informatics to support medical practice and clinical research [25]. With the success of the human genome project [11] and the rise of bioinformatics, many observers [12], [13], [26] have argued that the name medical informatics should be replaced with *biomedical informatics* to reflect the broad range of issues in biomedical research, clinical practice, and health-related applications.

Biomedical informatics is thus an interdisciplinary science that involves the incorporation of knowledge from diverse disciplines, including health science

T.-h. Kim et al. (Eds.): DTA/BSBT 2011, CCIS 258, pp. 133–142, 2011.

(e.g., medicine, dentistry, pharmacy, nursing), computer science, engineering, information science, cognitive science, biostatistics and mathematics. This emerging field encompasses scientific endeavors ranging from theoretical model construction to the building and evaluation of practical tools to solve complex problems in prevention and treatment of diseases, clinical/medical decision making, and delivery of effective health care. The focus of this paper is to propose a new methodology in developing a computer-assisted decision support system based on first-order logic to improve medical practice. The main contribution of our work is the systematic process of deriving knowledge as a data model from existing patient records. In addition to data modeling, the inference system of derived knowledge for decision support can be created automatically as well.

The presentation of our work is organized as follows. We review related work in medical decision support system in Section 2. Then, we discuss the system design and algorithms for knowledge inducing and inferring in Section 3. The system implementation and its performance evaluation are in Section 4. We conclude our work and discuss future research directions in Section 5.

2   Related Work

The automated learning of models from patient data and biomedical records has become more and more essential since the extensive computerization in healthcare industry and the significant advancement in genomic and proteomic technologies during the last decade. Medical and clinical databases have been created and constantly growing at an exponential rate. The development of an automatic and intelligent data analysis tool is an obvious solution to the data-flooding problem in medical domains [9], [18], [23], [27]. In recent years we have witnessed increasing number of applications on knowledge mining from biomedicine, clinical and health data. Roddick et al. [20] discussed the two categories of mining techniques applied over medical data: explanatory and exploratory. Explanatory mining refers to techniques that are used for the purpose of confirmation or making decisions. Exploratory mining is data investigation normally done at an early stage of data analysis in which an exact mining objective has not yet been set.

Explanatory mining in medical data has been extensively studied in the past decade employing various learning techniques. Bojarczuk et al. [1] applied genetic programming method to discover classification rules from medical data sets. Ghazavi and Liao [6] proposed the idea of fuzzy modeling on selected features medical data. Huang et al [8] introduced a system to apply mining techniques to discover rules from health examination data. Then they employed a case-based reasoning to support the chronic disease diagnosis and treatments. The recent work of Zhuang et al. [29] also combined mining with case-based reasoning, but applied a different mining method. Biomedical discovery support systems are recently proposed by a number of researchers [2], [3]. Some work [21] extended medical databases to the level of data warehouses.

Exploratory, as oppose to explanatory, is rarely applied to medical domains. Among the rare cases, Nguyen, Ho and Kawasaki [16] introduced knowledge visualization in the study of hepatitis patients. Palaniappan and Ling [17] applied the functionality of OLAP tools to improve visualization.

It can be seen from the literature that most medical knowledge discovery systems have been designed up to the stage of knowledge mining without further discussion on the final stage knowledge inferring and deployment. Kumar et al. [10] include decision-making unit with no detail regarding implementation in their decision-support system. Horng et al. [7] propose an expert system to classify microarray gene expression emphasizing only the gene selection and classification stages. The work of Exarchos et al. [4] is closely related to ours, but their methodology on the automatic creation of expert system is based on the fuzzy set. Our work, on the contrary, is a rule-based inductive expert system in that the knowledge is induced rules and the inference engine is also automatically generated from those rules. Uncertainty of knowledge applicability is based on the probabilistic concept.

3 System Architecture and Methodology

Electronic medical data are valuable resources for the automatic learning of useful knowledge to support scientific decision-making. Medical knowledge mining is an emerging area of computational intelligence applied to automatically analyze electronic medical records and health databases. The non-hypothesis driven analysis approach of data mining technology can induce knowledge from clinical data repositories and health databases. Various data mining methods have been proposed to learn useful knowledge from medical data, but major techniques adopted by many researchers are rule induction and classification tree generation [4], [7], [10], [28].

Our design of a knowledge induction system (Figure 1) is also based on a decision-tree induction concept. Decision tree induction [19] is a popular method for mining knowledge from data and representing the result as a classifier tree. Popularity is due to the fact that mining result in a form of decision tree is interpretability, which is more concern among casual users than a sophisticated method but lack of understandability. A decision tree is a hierarchical structure with each node contains decision attribute and node branches corresponding to different attribute values of the decision node. The goal of building decision tree is to partition data with mixing classes down the tree until each leaf node contains data with pure class.

In our system framework, we increase interpretability of the knowledge mining results by transforming the decision tree structure into a small set of decision rules. After a complete decision tree has been created, we calculate the probability of case occurrence augmented with each leaf node. In the phase of decision rule generation, these probability values will be sorted. Rules within the top ranking part will be displayed to assist medical practitioner for making decision. In the designed framework, probabilistic knowledge induction system is composed of four main components: data integration, tree induction, probabilistic-rule generation, and the knowledge inferring and answering engines. Data integration component is responsible for collecting data from different sources, cleaning and format transforming. These data are to be used by the tree induction component.

In order to build a decision tree, we need to choose the best attribute that contributes the most towards partitioning data to the purity groups. The metric to measure attribute's ability to partition data into pure class is *Info*, which is the number of bits required to encode a data mixture. To choose the best attribute we have to calculate information gain, which is the yield we obtained from choosing that attribute. The gain value of each candidate attribute is calculated. The attribute with maximum gain value is chosen to be the decision node. The process of data partitioning continues until the data subset along each tree branch has the same class label.

Given the induced tree, the probabilistic-rule generation component traverses each tree branch to calculate the likelihood of path occurrence. This likelihood is interpreted as the probability of event and associated to the rule generated from the path traversal. The generated probabilistic rules are then sorted. Rules at the top ranking (specified by the given minimum probability) are stored in the knowledge base as the probabilistic knowledge and could be used for recommendation or answering query to the medical practitioner. Algorithms for knowledge induction based on tree structure (Algorithm 1), probabilistic-rule generation from decision tree (Algorithm 2), and probabilistic knowledge inferring to answer the most probable class decision on new case (Algorithm 3) are given in Figures 2, 3, and 4, respectively. The induced probabilistic knowledge base is a major part of our medical inductive expert system. It is a rule-based expert system with two important automated components: automatic knowledge acquisition subsystem and the inference engine to support explanation and reasoning.

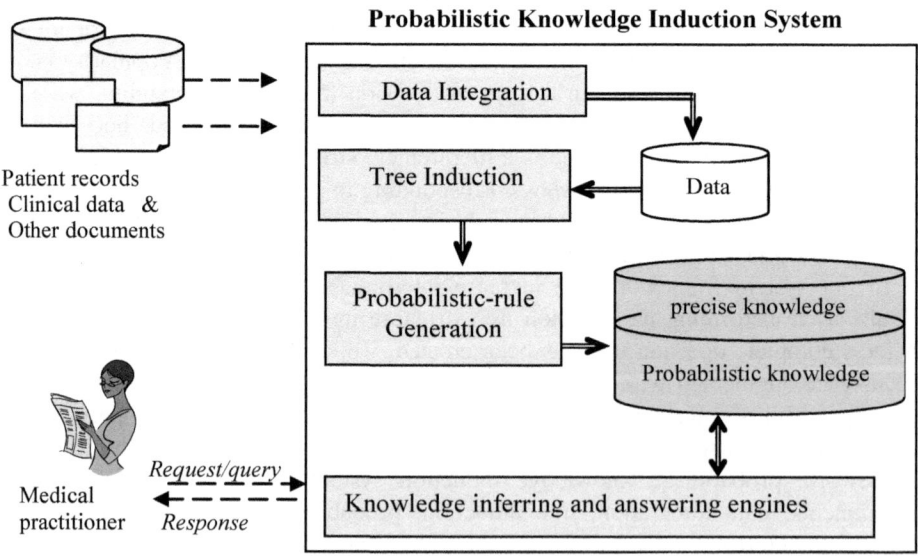

Fig. 1. A framework for probabilistic knowledge induction based on the decision tree

Algorithm 1. Knowledge induction
        Input: a data set formatted as Prolog clauses
        Output: a decision tree with node and edge structures
(1)   Initialization
(1.1)   Clear temporary knowledge base (KB)
(1.2)   Set node counter = 0
(1.3)   Scan data set to get information about data attributes and instances
(2)   Building tree
(2.1)   Increment node counter
(2.2)   Repeat steps 2.2.1-2.2.4 until there is no more attributes left for creating
                decision nodes
(2.2.1)   Compute Info value of each candidate attribute
(2.2.2)   Choose the attribute that yields minimum Info to be decision node
(2.2.3)   Assert edge and node information into KB
(2.2.4)   Split data instances along node branches
(2.3)   Repeat steps 2.1 and 2.2 until the lists of positive and negative instances
                are empty
(2.4)   Output tree structure containing node and edge predicates

Fig. 2. Knowledge induction algorithm to generate a decision-tree structure

Algorithm 2. Probabilistic knowledge generation

    Input: a decision tree with node and edge structures, and a probability
                threshold
    Output: a set of probabilistic rules ranking from the highest probability
(1)   Traverse tree from a root node to each leaf node
(1.1)   Collect edge information and count number of data instances
(1.2)   Compute probability as a proportion
                (number of instances at leaf node) / (total data instances in a data set)
(1.3)   Assert a rule containing a triplet (attribute-value pair, class, probability
                value) into KB
(2)   Sort rules in the KB in descending order according to the rules' probability
(3)   Remove rules that have probability less than the specified threshold
(4)   Assert selected rules into the KB and return KB as an output

Fig. 3. An algorithm to generate probabilistic decision rules from the tree structure

Algorithm 3. Probabilistic knowledge inferring
    Input: a KB containing probabilistic knowledge, and a new case with
                unknown class value
    Output: a decision on most likely class of the new case
(1)   Read all attribute-value pairs appeared in the given case
(2)   Compare the pairs with relevant rules in the KB to get the decision class value
(3)   Compute the decision confidence as
                (number of matched attribute-value pair) × (probability of the decision rule)
(4)   Output a final decision based on the voting scheme

Fig. 4. A utilization of probabilistic decision rules to predict unknown class

4   Implementation and Performance Evaluation

The implementation of a probabilistic knowledge induction component and the rule-based inference engine is based on a logic programming paradigm. A rapid prototype of the proposed inductive expert system is provided in a declarative style using second-order Horn clauses [15]. Prolog code appeared in appendices follows the syntax of SWI Prolog (www.swi-prolog.org). The intuitive idea of our design and implementation is that for such a complicated knowledge-base system coding should be done declaratively at a high level to alleviate the burden of programmers. The advantages of declarative style are thus a decrease in program development time and the increase in expressiveness of knowledge representation and efficiency of knowledge utilization.

Data Format. In logic programming, program and data take the same format, i.e. all are in Prolog clausal form. For the purpose of demonstration, we use the health examination data of 86 patients for discharge decision after their operations. Each patient record contains eight observed attributes. The general conditions such as blood pressure and temperature are observed to determine whether the patient is in good condition and should be sent home shortly (class=home), or the condition is quite moderate and should stay at the hospital ward for further follow up (class=ward). The post-operative data set, which is downloadable from the UCI repository [5], in Prolog clausal form is shown some part as the following

```
attribute(internalTemp,      [mid, high, low]).
attribute(surfaceTemp,       [mid, high, low]).
attribute(oxygenSaturation,  [excellent, good, fair, poor]).
attribute(bloodPressure,     [high, mid, low]).
attribute(tempStability,     [stable, mod-stable, unstable]).
attribute(coreTempStability, [stable, mod-stable, unstable]).
attribute(bpStability,       [stable, mod-stable, unstable]).
attribute(comfort,           [5,7,10,15]).
attribute(class,        [home, ward]).
instance(1,class=ward,[internalTemp=mid, surfaceTemp=low,
            oxygenSaturation=excellent, bloodPressure=mid,
            tempStability=stable, coreTempStability=stable,
            bpStability=stable, comfort=15]).
instance(2,class=home,[internalTemp=mid, surfaceTemp=high,
            oxygenSaturation=excellent, bloodPressure=high,
            tempStability=stable, coreTempStability=stable,
            bpStability=stable, comfort=10]).
```

Knowledge Induction and Probabilistic Rule Generation. The three algorithms (explained in the previous section) are called by the main module, which is the top-level of our program implementation. The Prolog coding of main module is as follows:

```
main :- init(AllAttr,EdgeList), getnode(N),
        create_edge_onelevel(N,AllAttr,EdgeList),
        addKnowledge,  write(chooseMinProb), read(Min),
        selectRule(Min,Res), maplist(writeln,Res).
```

The predicates init and getnode initialize the tree structure. The tree-based knowledge induction process starts when the main module invokes the predicate create_edge_onelevel to build a decision tree one level at a time. After the complete tree structures are created, the predicates addKnowledge and selectRule are invoked to compute probability along each tree branch to generate probabilistic rules, and then select only rules that could occur at the probability level higher than the specified threshold. User can control minimum probability level through the interactive interface of the system (Figure 5).

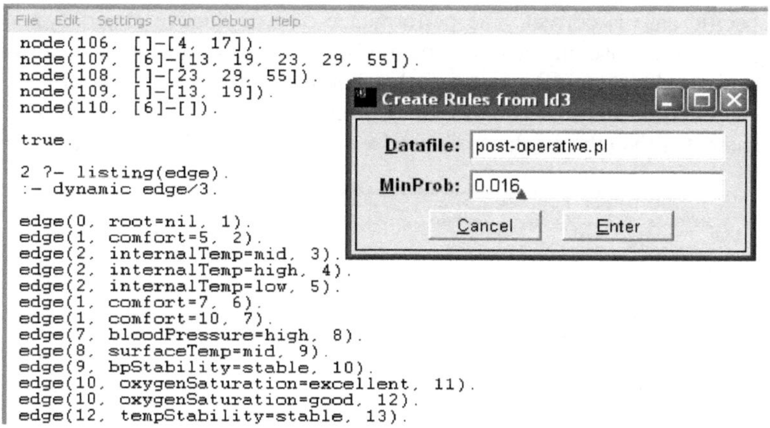

Fig. 5. Interactive user interface and the generated tree structures

Automatic Inference Engine and Knowledge Base Creation. The probabilistic rules are then written in file to serve as a knowledge base of the inductive expert system. A rule-based inference engine is also automatically created as shown in Figure 6.

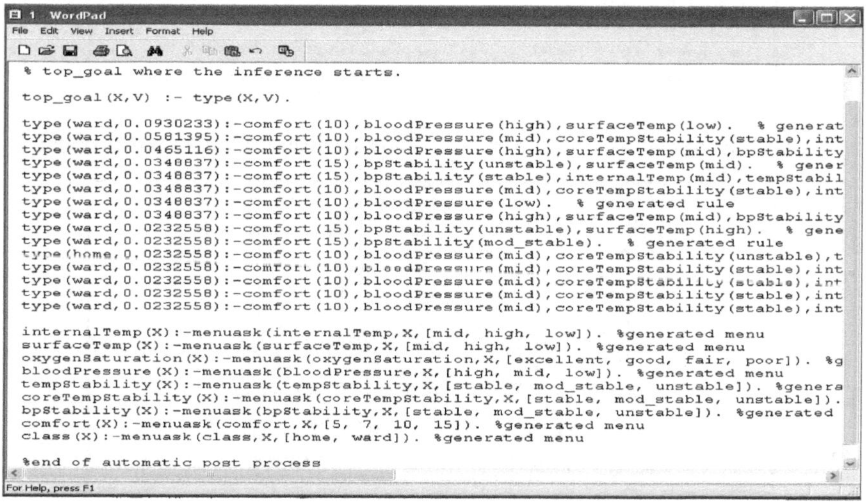

Fig. 6. Knowledge base contents and a rule-based inference engine

Performance Evaluation. We evaluate correctness of the induced probabilistic knowledge by dividing data set into two subsets: a training set containing 70 patient records, and a test set containing 16 patient records. A training set is used in the knowledge induction phase. The training results, which are knowledge base and inference rules, are then tested by the test set. Figure 7 illustrates the test process of data instance number 71 by the expert system shell. The recommendation given by the inductive expert system is that the patient should be sent to the general ward with probability (or confidential level) 0.0348837. The actual diagnosis made by the doctor is also admission to the general ward. Therefore, the recommendation given by the inductive expert system for the specific case is correct. The performance of an inductive expert system is also tested against other machine learning methods: ID3, Prism, and Neural network. The experimental result (Figure 8) confirms that our inductive expert system can generate probabilistic decision rules as good as the data model obtained from the Neural network method, and better than the results produced by the ID3 and the Prism methods.

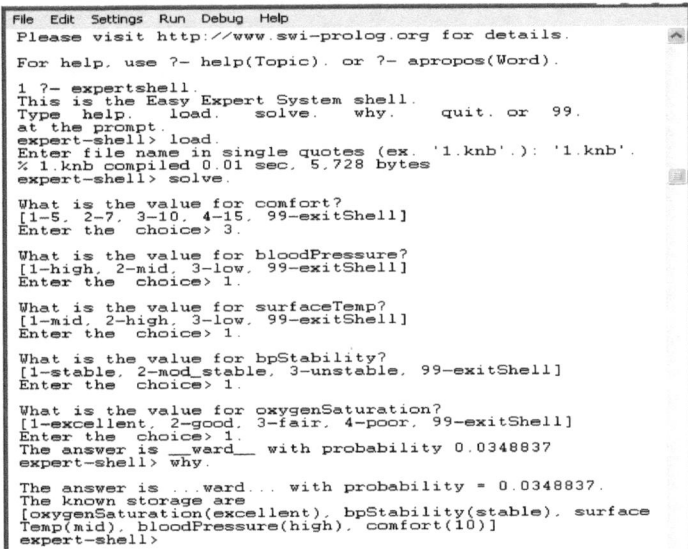

Fig. 7. The process of correctness testing of the automatic inductive expert system

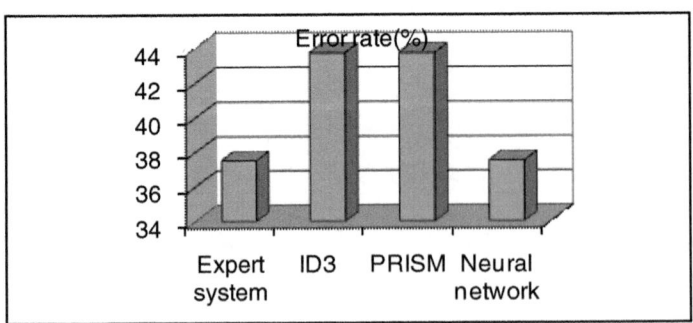

Fig. 8. Comparison on prediction error rate of the inductive expert system against the ID3, Prism, and Neural network methods

5 Conclusion

Biomedical informatics deals with biomedical information, its structure, acquisition and optimal use for problem solving and decision making. Medical knowledge discovery is a research area that employing machine learning techniques to acquire knowledge from health examination and clinical data. Knowledge extraction from huge amount of health databases is expected to ease the medical decision-making process. The ultimate goal of knowledge extraction is to generate the most accurate and useful knowledge and represent it in an understandable format. Such goal is, however, difficult to accomplish due to the learning complexity of knowledge induction methods and the nonconformity of the database contents. Most of the time knowledge discovery from medical databases results in reporting large number of irrelevant knowledge. We thus focus our study on this issue and devise a technique to extract a limited number of knowledge that is most likely relevant to the specific domain.

In medical domains, interpretability of results is an important feature of the data analysis tool. Medical practitioners need a system that can produce accurate results in an understandable form. Therefore, knowledge represented as rules has been widely used for knowledge discovery in medical applications. Nevertheless, in medical applications learning techniques tend to generate a lot of rules. Too many rules, some are redundant and uninteresting, cause problems to the medical practitioners because a truly relevant one can be easily overlooked. We thus propose a rule induction method based on the decision-tree structure that adopts the probability concept to select the most probable applicable rules. The selected rules are then automatically transformed to be the knowledge and the inference engine in the medical expert system to support decision making. The experimentation on our inductive expert system confirms a good performance of the system on recommending patient discharge decision after an operation. Direct application of medical probabilistic knowledge base is for medical related decision-making. Other indirect but obvious application of such knowledge is to pre-process other data sets by grouping it into focused subset containing only relevant data instances.

Acknowledgments. This work has been supported by grants from the National Research Council of Thailand (NRCT) and Suranaree University of Technology via the funding of Data Engineering Research Unit.

References

1. Bojarczuk, C., et al.: A constrained-based syntax genetic programming system for discovering classification rules: Application to medical data sets. Artificial Intelligence in Medicine 30, 27–48 (2004)
2. Bratsas, C., et al.: KnowBaSICS-M: An ontology-based system for semantic management of medical problems and computerized algorithmic solutions. Computer Methods and Programs in Biomedicine 83, 39–51 (2007)
3. Correia, R., Kon, F., Kon, R.: Borboleta: A mobile telehealth system for primary homecare. In: Proc. ACM Symposium on Applied Computing, pp. 1343–1347 (2008)
4. Exarchos, T.P., et al.: A methodology for the automatic creation of fuzzy expert systems for ischaemic and arrhythmic beat classification based on a set of rules obtained by a decision tree. Artificial Intelligence in Medicine 40, 187–200 (2007)

5. Frank, A., Asuncion, A.: UCI Machine Learning Repository. University of California, School of Information and Computer Science, Irvine (2010), http://archive.ics.uci.edu/ml
6. Ghazavi, S., Liao, T.: Medical data mining by fuzzy modeling with selected features. Artificial Intelligence in Medicine 43(3), 195–206 (2008)
7. Horng, J., et al.: An expert system to classify microarray gene expression data using gene selection by decision tree. Expert Systems with Applications 36, 9072–9081 (2009)
8. Huang, M., Chen, M., Lee, S.: Integrating data mining with case-based reasoning for chronic disease prognosis and diagnosis. Expert Systems with Applications 32, 856–867 (2007)
9. Kononenko, I.: Machine learning for medical diagnosis: History, state of the art and perspective. Artificial Intelligence in Medicine 1, 89–109 (2001)
10. Kumar, K.A., Singh, Y., Sanyal, S.: Hybrid approach using case-based reasoning and rule-based reasoning for domain independent clinical decision support in ICU.Expert Systems with Applications 36, 65-71 (2009)
11. Lander, E.S., et al.: Initial sequencing and analysis of the human genome. Nature 409, 860–921 (2001)
12. Maojo, V., Kulikowski, C.: Bioinformatics and medical informatics: Collaborations on the road to genomic medicine? J. American Medical Informatics Association 10(6), 515–522 (2003)
13. Maojo, V., et al.: Medical informatics and bioinformatics: European efforts to facilitate synergy. J. Biomedical Informatics 34, 423–427 (2001)
14. Miller, R.A., Pople, H.E., Myers, J.D.: INTERNIST-1, An experimental computer-based diagnostic consultant for general internal medicine. New England J. Medicine 307(8), 468–476 (1982)
15. Nadathur, G., Miller, D.: Higher-order Horn clauses. J. ACM 37, 777–814 (1990)
16. Nguyen, D., Ho, T., Kawasaki, S.: Knowledge visualization in hepatitis study. In: Proc. Asia-Pacific Symposium on Information Visualization, pp. 59–62 (2006)
17. Palaniappan, S., Ling, C.: Clinical decision support using OLAP with data mining. Int. J. Computer Science and Network Security 8(9), 290–296 (2008)
18. Pandey, B., Mishra, R.B.: Knowledge and intelligent computing system in medicine. Computers in Biology and Medicine 39, 215–230 (2009)
19. Quinlan, J.R.: Induction of decision trees. Machine Learning 1, 81–106 (1986)
20. Roddick, J., Fule, P., Graco, W.: Exploratory medical knowledge discovery: Experiences and issues. ACM SIGKDD Explorations Newsletter 5(1), 94–99 (2003)
21. Sahama, T., Croll, P.: A data warehouse architecture for clinical data warehousing. In: Proc. 12th Australasian Symposium on ACSW Frontiers, pp. 227–232 (2007)
22. Schwartz, W.B.: Medicine and the computer: The promise and problems of changes. New England J. Medicine 283, 1257–1264 (1970)
23. Shillabeer, A., Roddick, J.F.: Establishing a lineage for medical knowledge discovery. In: Proc. 6th Australasian Conf Data Mining and Analytics, pp. 29–37 (2007)
24. Shortliffe, E.H.: Computer-based medical consultations: MYCIN. Elsevier (1976)
25. Shortliffe, E.H., Cimino, J.J.: Biomedical informatics: Computer applications in health care and biomedicine, 3rd edn. Springer, Heidelberg (2006)
26. Stead, W.: The challenge of bridging between disciplines. J. American Medical Informatics Association 8(1), 105 (2001)
27. Truemper, K.: Design of logic-based intelligent systems. John Wiley & Sons (2004)
28. Zhou, Z., Jiang, Y.: Medical diagnosis with C4.5 rule preceded by artificial neural network ensemble. IEEE Trans. Information Technology in Biomedicine 1, 37–42 (2003)
29. Zhuang, Z., Churilov, L., Burstein, F.: Combining data mining and case-based reasoning for intelligent decision support for pathology ordering by general practitioners. European J. Operational Research 195(3), 662–675 (2009)

Method for Protein Active Sites Detection Based on Fuzzy Decision Trees

Georgina Mirceva, Andreja Naumoski, Viktorija Stojkovik,
Damjan Temelkovski, and Danco Davcev

Ss. Cyril and Methodius University in Skopje, Macedonia
{georginamirceva,andrejanaumoski,damjantemelkovski,
dancodavcev}@gmail.com, viki_stojkovik@yahoo.com

Abstract. The knowledge of the protein functions is very important in the development of new drugs. Many experimental methods for determining protein function exist, but due to their complexity the number of protein structures with unknown functions is rapidly growing. So, there is an obvious necessity for development of computer methods for annotating protein structures. In this paper we present a fuzzy decision tree based method for protein active sites detection, which could be used for annotating protein structures. We extract several features of the amino acids, and then using different membership functions we build fuzzy decision trees in order to detect the possible active sites. We provide some experimental results of the evaluation of our method. Additionally, our method is compared with several existing methods for protein active sites detection.

Keywords: Protein active site, Fuzzy Decision Tree.

1 Introduction

Proteins are one of the most important compounds in human cells since they are involved in many organisms' process. Structural genomics projects discover protein structures with unknown function. The knowledge of the protein functions is crucial for development of new drugs, better crops and synthetic biochemical. Biologists require accurate and automated methods for predicting protein functions in reasonable time. Experimental methods for determining protein function exist, but they are expensive and complex. As a result of this, the number of protein structures with unknown functions is rapidly growing. Therefore, there is an obvious necessity for development of computer methods for protein function annotation.

There are several types of methods for discovering protein functions based on the information that they consider. The protein functions can be determined by analysis of the structural and sequence homology [1] of the protein structures, but in this way only a global similarity can be determined. Protein functions can also be determined by analysis of protein-protein interaction networks [2], [3]. These methods consider only information about which proteins interact each other, but don't take into account their sequence and structure. They also require a priori knowledge about which structures it

T.-h. Kim et al. (Eds.): DTA/BSBT 2011, CCIS 258, pp. 143–150, 2011.

interacts with. However, the acquisition of this knowledge is very expensive. Other methods analyse the conservation of the protein sequence and structure [4]. They are based on the fact that some part of the protein sequence/structure is stable during evolution. The conservation of the protein sequence/structure could be determined by multiple-alignment of protein sequences/structures [5]. The most relevant methods for annotating protein structures are based on detecting protein active sites [6] since molecular biologist manually annotate protein structures in similar manner. For example, the knowledge stored in the BIND database [7] is obtained experimentally by detecting protein active sites. Therefore, in this research we focus our efforts on protein active sites detection.

The protein active sites are the amino acids where protein structures interact with each other. For protein active sites detection many features of amino acid residues could be considered, i.e. sequence and structure conservation [4], Accessible Surface Area (ASA) [8], depth index [9], protrusion index [10], hydrophobicity [11], and other physicochemical and shape features. Then, the most relevant features of the active sites could be considered and the possible protein interactions could be predicted. No single feature absolutely differentiates the protein active sites from the other amino acids found at the protein surface. The methods for predicting protein interactions could be based on many criteria like change in ASA [12], Van der Walls distance [13], physicochemical and shape complementarity [14] etc. These methods should determine where the interaction would occur, as well as what are the corresponding protein functions for the interacting protein structures.

There are many methods for protein active site detection [13], [15], [16], [17], [18], [19]. Owing to the popularity of this issue, there are a lot of available tools and web servers for analysis of protein-protein interactions. Nevertheless, these methods are very sensitive to small changes in the data. Consequently, the small changes of the amino acids' features obtained during evolution could significantly influence the predictions. In order to provide resistance to small changes in the data, we use a fuzzy decision tree [20] based method for protein active sites detection.

The robustness of data changes and resistance to over-fitting of the fuzzy inducted trees is the main reason for extensive research on fuzzy set based machine learning. Wang and Mendel [21] have presented an algorithm for generating fuzzy rules by learning from examples. Inspired by the classical decision trees proposed by Quinlan [22], there are substantial works on fuzzy decision trees. Janikow [20], Olaru and Wehenkel [23] have presented different fuzzy decision tree induction methods. Suárez and Lutsko [24], and Wang et al. [25] have presented optimizations of fuzzy decision trees. The fuzzy decision trees, their optimizations, advantages and disadvantages over classical decision trees are studied in [26].

In this paper, we present a method for protein active sites detection based on fuzzy decision trees. We extract several features of the amino acid residues, and then we induce fuzzy decision trees using different membership functions in order to detect the protein active sites. We provide some experimental results of the evaluation of our method. Also, the method is compared with several existing methods for protein active sites detection.

Our research method is presented in section 2. In section 3 some experimental results are given, while section 4 concludes the paper.

2 Our Method

In this paper we present a method for protein active sites detection. First, we extract several features of the amino acid residues of the protein structure and thus forming a feature vector for each amino acid. Then, we induce fuzzy decision trees that are later used for protein active sites detection.

2.1 Extraction of the Features of the Amino Acids

In this research we consider four amino acids' features: accessible surface area, depth index, protrusion index and hydrophobicity. The accessible surface area (ASA) is first described by Lee and Richards [27] and is typically calculated using the 'rolling ball' algorithm [8] developed by Shrake and Rupley, which uses a sphere of a particular radius to 'probe' the surface of the molecule. The ASA is defined as the area on the surface of the sphere of radius R, which can be placed in contact with the probe. The radius R is given by the sum of the van der Waals radius of the atom and the chosen radius of the solvent molecule (probe). Typically, the solvent molecule has the same radius (1.4 Å) as the water molecule.

According to the ASA value we estimate whether a given amino acid is at the protein surface or it is buried in its interior. We consider a given amino acid to be at the surface if at least 5% of its area could be solved by the rolling probe [28]. We use the values for the total area of the amino acids given in [28]. At the second stage of the method, we consider only the amino acids that are located at the protein surface since the amino acids buried in the interior could not be active sites.

The depth of an atom (DPX) [9] is defined as its distance (Å) to the closest solvent accessible atom. The depth index is thus zero for the solvent accessible atoms, and greater than zero for atoms buried in the protein interior, with deeply buried atoms having higher DPX values [9].

The algorithm for calculating the protrusion index [10] is as follows. For each non-hydrogen atom, the number of the heavy atoms within a given distance R is calculated. We have used $R=10$ Å as suggested in [10]. The number of atoms within the sphere is multiplied by the mean atomic volume (20.1 ± 0.9 Å), which gives the volume occupied by a protein within a sphere V_{int}. The remaining volume of the sphere V_{ext} is calculated as the difference between the volume of the sphere and V_{int}. Finally, the protrusion index CX is thus calculated as $CX=V_{ext}/V_{int}$ [10]. For a given amino acid residue, both DPX and CX are calculated as the mean depth/protrusion index of all atoms in that amino acid.

The hydrophobicity of an amino acid is a number that presents its hydrophobic properties. This feature is very important because hydrophobic amino acids tend to be internal, while hydrophilic amino acids are usually found towards the protein surface. There are several scales for amino acids' hydrophobicity. In this research we have used the Kyte and Doolittle scale [11].

2.2 Fuzzy Decision Trees

The induction of fuzzy decision trees is similar to the induction of classical decision trees [22] with modified induction criteria [20]. The potential of fuzzy decision trees in improving the robustness and generalization in classification is due to the use of fuzzy reasoning. Using crisp discretization, the decision space is partitioned into non-overlapping subspaces where each example is assigned to a single class. On the other hand, a fuzzy decision tree gives results within [0, 1] as possibility that an example belongs to a given class. Therefore, the fuzzy decision trees provide a robust way to avoid misclassifications.

In the induction of classical decision trees, a suitable threshold T that discretizes a continuous attribute A into two intervals $A_1 = [min(A), T]$ and $A_2 = (T, max(A)]$ is determined based on the information gain.

The soft discretization could be viewed as an extension of the hard discretization where the information measures defined in the probability domain are extended to new definitions in the possibility domain. A crisp set A_c is expressed with a sharp characterization function $A_c(a)$: $\Omega \rightarrow [0, 1]$; $a \in \Omega$, alternatively a fuzzy set A is characterized with a membership function $A(a)$: $\Omega \rightarrow [0, 1]$; $a \in \Omega$. The probability of fuzzy set A is defined as $P_F(A) = \int_{\Omega} A(a)dP$, where dP is a probability measure on Ω, while F denotes an associated fuzzy term. If A is defined on discrete domain and $P(a_i)=p_i$, then $P_F(A) = \sum_{i=1}^{m} A(a_i)p_i$. Let $Q=\{A_1,...A_k\}$ be a family of fuzzy sets on Ω. Q is called a fuzzy partition of Ω if $\sum_{i=1}^{k} A_i(a) = 1, \forall a \in \Omega$. The soft discretization is defined with the cross point T and the membership functions of the fuzzy set pair A_1 and A_2 where $A_1(a) + A_2(a) = 1$. The cross point T is determined based on whether it can maximize the information gain, and the membership functions of the fuzzy set pair are determined according to the characteristics of the data. The fuzzy class entropy in partition S_i is defined by (1), where $p(c_j, S_i)$ is the fuzzy proportion of examples in S_i.

$$E_F(S_i) = -\sum_{j=1}^{k} p(c_j, S_i)\log(p(c_j, S_i))$$

$$p(c_j, S_i) = N_F^{S_i c_j} / N_F^{S_i} , i = 1,2$$

$$N_F^{S_i c_j} = \sum_{a_k \in c_j} A_i(a_k) , i = 1,2 \tag{1}$$

$$N_F^{S_i} = \sum_{k=1}^{|S|} A_i(a_k) , i = 1,2$$

After soft discretization, the class information entropy is calculated with the probability of fuzzy partition using (2).

$$E_F(A,T,S) = \frac{N_F^{S_1}}{N_F^{S}} E_F(S_1) + \frac{N_F^{S_2}}{N_F^{S}} E_F(S_2)$$

$$N_F^{S} = \sum_{i=1}^{|S|} (A_1(a_i) + A_2(a_i)) \tag{2}$$

Similar to classical decision tree induction, the information gain $E_F(S) - E_F(A, T, S)$ is used to generate the best discretization for the corresponding attribute.

The fuzzy decision tree induction includes eight steps as follows:

1: Assume that the current node denoted as S contains N examples. Sort the examples according to the values of the attribute A;

2: Generate candidate cut points T using the class boundary points;

3: Fuzzify the cut points to generate candidate soft discretizations using the fuzzy set pair A_1 and A_2, which form a fuzzy partition on the cut point;

4: Evaluate each candidate;

5: Select the soft discretization having a minimum value of E_F for the attribute A;

6: Repeat steps 1-5 for the other attributes;

7: Select the attribute A_j whose discretization has minimal $E_F(A, T, S)$ to generate two child branches and nodes;

8: Calculate the truth level for each of the two branches: $\eta_1 = \dfrac{N_F^{S_1}}{N_F^{S}}, \eta_2 = \dfrac{N_F^{S_2}}{N_F^{S}}$.

If $\eta_1 \leq \alpha$ or $\eta_2 \leq \alpha$, then delete the corresponding branches. If $\eta_1 > \alpha$ or $\eta_2 > \alpha$, then calculate the truth level of each branch belonging to the j-th class:

$$\mu_{1,j} = \frac{\sum_{a_i \in c_j} A_1(a_i)}{N_F^{S_1}}, \quad \mu_{2,j} = \frac{\sum_{a_i \in c_j} A_2(a_i)}{N_F^{S_2}}.$$

If $\max_{j=1}^{k} \mu_{1,j} \geq \beta$ or $\max_{j=1}^{k} \mu_{2,j} \geq \beta$ then the corresponding branch is terminated as a leaf, and this leaf is assigned as the class c_j. Otherwise, S is partitioned into $S_1 = \{s \mid A_1(a_i) \geq \lambda, a_i \in S\}$ and $S_2 = \{s \mid A_2(a_i) \geq \lambda, a_i \in S\}$ and the above steps are repeated for each child node until the above criteria are satisfied. In this research we used $\alpha = 0.1$, $\beta = 0.8$ and $\lambda = 0.5$.

3 Experimental Results

For evaluation of the proposed method we used a part of the BIND database [7] that contains experimentally obtained knowledge about protein active sites. We used a representative training dataset so that each pair of protein chains has less than 40% sequence similarity using the selection criterion in [30]. Since this selection criterion considers protein chains with low sequence similarity, representative protein chains are considered as training data. In this way we obtained 1062 training protein chains with 365862 amino acids in total from which 284168 are on the protein surface. From these surface amino acids, 26889 are active sites according to the BIND database. Since the non-active class is dominant, we balance the training dataset in that way that the active amino acids are considered several times until the training dataset is balanced. We want to note that the balancing is done only on the training data in order to avoid preferring the non-active class. In this way we obtained 514550 training amino acids. We have chosen 3980 protein chains in the test set, thus obtaining 812679 surface amino acids from which 83884 are active sites according to the BIND database.

For evaluation, we used the Area under ROC curve (AUC-ROC) as efficiency measure. In order to calculate this measure, we have to calculate TP (true positives), FP (false positives), TN (true negatives) and FN (false negatives), where positive examples are those examples that are classified as active sites and negative examples are those examples that are classified in the non-active class. TP is the number of the correctly classified positive examples, FP is the number of negative examples classified as positive, TN is the number of correctly classified negative examples and FN is the number of positive examples classified as negative. Based on these measures, the True Positive Rate (TPR), the True Negative Rate (TNR) and the Area under ROC curve (AUC-ROC) could be calculated (3). The AUC-ROC is the most appropriate measure, especially when the classes are of very different sizes, like it is in our case. AUC-ROC is a value between 0 and 1, where 1 represents a perfect prediction, while 0 an inverse prediction.

$$
\begin{aligned}
&\text{TPR} = \text{TP}/(\text{TP} + \text{FN}) \\
&\text{TNR} = \text{TN}/(\text{TN} + \text{FP}) \\
&\text{AUC} - \text{ROC} = \text{TPR} * \text{TNR} + \text{TPR} * (1 - \text{TNR})/2 + \text{TNR}(1 - \text{TPR})/2 \\
&\qquad\qquad = (\text{TPR} + \text{TNR})/2
\end{aligned}
\tag{3}
$$

First, we examined the influence of the number and type of membership functions used in the induction of fuzzy decision trees. We examined the triangular, trapezoidal and Gaussian membership functions (MFs) [29] using N = 3, 4, 5 and 10 MFs. The results are given in Table 1.

Table 1. The results obtained by our method using different types and different number N of membership functions.

	$N=3$	$N=4$	$N=5$	$N=10$
Triangular	0.5472	0.5623	0.5735	0.5741
Trapezoidal	0.5369	0.5473	0.5623	0.5736
Gaussian	0.5472	0.5665	0.5754	0.5743

As it can be seen from Table 1, as the number of membership functions N increases, also the AUC-ROC increases. Using Gaussian MFs highest AUC-ROC is obtained, while the AUC-ROC obtained using trapezoidal MFs is insignificantly lower than by the other MFs. According to Table 1 we can conclude that using 5 Gaussian MFs we obtain the most accurate models. We want to note that N should not be set too large since the induction of the tree would be very slow and also the model could be over-fitted.

Next, we compared the proposed method with several existing methods for detecting protein active sites. In this analysis PPI-PRED [16], PDBeMotif [31] and PRISM [18] methods were used. PDBeMotif obtained lowest AUC-ROC (0.5015), then PPI-PRED method follows (0.5916), while PRISM obtained highest AUC-ROC (0.6541). This analysis showed that our method is better than PDBeMotif. PPI-PRED method obtained higher AUC-ROC than our method, but it have to be noticed that the decision with our method is made in several seconds, while PPI-PRED takes several minutes. PRISM method obtained highest AUC-ROC, but it is significantly slower than the other methods since it perform exhaustive structural alignment of the protein structure with template structures.

In this research we used the basic membership functions. We suppose that using more sophisticated membership functions, like sigmoidal, log-normal and bell, we can significantly increase the prediction power of our method.

4 Conclusion

In this paper, we presented our fuzzy decision tree based method for protein active site detection. Several features of the amino acids were extracted, and then fuzzy decision trees were induced using different types and different number of membership functions.

The analysis showed that as the number of membership functions increases, the prediction power of the method increases. However, this number should not be too large since the induction of the tree would be very slow and also the model could be over-fitted. We have compared our method with several existing methods for protein active sites detection. Our method outperforms PDBeMotif. PPI-PRED and PRISM have greater prediction power than our method, but they are significantly slower.

Nevertheless, in this research we used the most basic membership functions. We suppose that using more sophisticated membership functions, like sigmoidal, log-normal and bell, we can significantly increase the prediction power of our method.

References

1. Todd, A.E., Orengo, C.A., Thornton, J.M.: Evolution of function in protein superfamilies, from a structural perspective. J. Mol. Biol. 307(4), 1113–1143 (2001)
2. Kirac, M., Ozsoyoglul, G., Yang, J.: Annotating proteins by mining protein interaction networks. Bioinformatics 22(14), e260–e270 (2006)
3. Sharan, R., Ulitsky, I., Shamir, R.: Network-based prediction of protein function. Mol. Sys. Bio. 3, 88 (2007)
4. Panchenko, A.R., Kondrashov, F., Bryant, S.: Prediction of functional sites by analysis of sequence and structure conservation. Protein Science 13(4), 884–892 (2004)
5. Leibowitz, N., Fligelman, Z.Y., Nussinov, R., Wolfson, H.J.: Automated multiple structure alignment and detection of a common substructure motif. Proteins 43(3), 235–245 (2001)
6. Tuncbag, N., Kar, G., Keskin, O., Gursoy, A., Nussinov, R.: A survey of available tools and web servers for analysis of protein-protein interactions and interfaces. Briefings in Bioinformatics 10(3), 217–232 (2009)
7. Bader, G.D., Donaldson, I., Wolting, C., Ouellette, B.F., Pawson, T., Hogue, C.W.: BIND: the Biomolecular Interaction Network Database. Nucleic Acids Res. 29(1), 242–245 (2001)
8. Shrake, A., Rupley, J.A.: Environment and exposure to solvent of protein atoms. Lysozyme and insulin. J. Mol. Biol. 79(2), 351–371 (1973)
9. Pintar, A., Carugo, O., Pongor, S.: DPX: for the analysis of the protein core. Bioinformatics 19(2), 313–314 (2003)
10. Pintar, A., Carugo, O., Pongor, S.: CX, an algorithm that identifies protruding atoms in proteins. Bioinformatics 18(7), 980–984 (2002)
11. Kyte, J., Doolittle, R.F.: A simple method for displaying the hydropathic character of a protein. J. Mol. Biol. 157(1), 105–132 (1982)
12. Jones, S., Thornton, J.M.: Analysis of protein-protein interaction sites using surface patches. J. Mol Biol. 272(1), 121–132 (1997)
13. Aytuna, A.S., Gursoy, A., Keskin, O.: Prediction of protein-protein interactions by combining structure and sequence conservation in protein interfaces. Bioinformatics 21(2), 2850–2855 (2005)
14. Lawrence, M.C., Colman, P.M.: Shape complementarity at protein/protein interfaces. J. Mol. Biol. 234(4), 946–950 (1993)

15. Neuvirth, H., Raz, R., Schreiber, G.: ProMate: a structure based prediction program to identify the location of protein-protein binding sites. J. Mol. Biol. 338(1), 181–199 (2004)
16. Bradford, J.R., Westhead, D.R.: Improved prediction of protein-protein binding sites using a support vector machines approach. Bioinformatics 21(8), 1487–1494 (2005)
17. Murakami, Y., Jones, S.: SHARP2: protein-protein interaction predictions using patch analysis. Bioinformatics 22(14), 1794–1795 (2006)
18. Ogmen, U., Keskin, O., Aytuna, A.S., Nussinov, R., Gursoy, A.: PRISM: protein interactions by structural matching. Nucleic. Acids. Res. 33(2), W331–W336 (2005)
19. Jones, S., Thornton, J.M.: Prediction of protein-protein interaction sites using patch analysis. J. Mol. Biol. 272(1), 133–143 (1997)
20. Janikow, C.Z.: Fuzzy decision trees: issues and methods. IEEE Transactions on Systems, Man, and Cybernetics 28(1), 1–14 (1998)
21. Wang, L.X., Mendel, J.M.: Generating fuzzy rules by learning from examples. IEEE Transactions on Systems, Man, and Cybernetics 22(6), 1414–1427 (1992)
22. Quinlan, R.J.: Decision trees and decision making. IEEE Transactions on Systems, Man, and Cybernetics 20(2), 339–346 (1990)
23. Olaru, C., Wehenkel, L.: A complete fuzzy decision tree technique. Fuzzy Sets and Systems 138(2), 221–254 (2003)
24. Suárez, A., Lutsko, J.F.: Globally optimal fuzzy decision trees for classification and regression. IEEE Transactions on Pattern Analysis and Machine Intelligence 21(12), 1297–1311 (1999)
25. Wang, X., Chen, B., Olan, G., Ye, F.: On the optimization of fuzzy decision trees. Fuzzy Sets and Systems 112(1), 117–125 (2000)
26. Chen, Y.-L., Wang, T., Wang, B.-S., Li, Z.-J.: A Survey of Fuzzy Decision Tree Classifier. Fuzzy Information and Engineering 1(2), 149–159 (2009)
27. Lee, B., Richards, F.M.: The interpretation of protein structures: Estimation of static accessibility. J. Mol. Biol. 55(3), 379–400 (1971)
28. Chothia, C.: The Nature of the Accessible and Buried Surfaces in Proteins. J. Mol. Biol. 105(1), 1–12 (1976)
29. Klir, G.J., Yuan, B.: Fuzzy sets and fuzzy logic: theory and applications, 1st edn. Prentice-Hall (1995)
30. Chandonia, J.-M., Hon, G., Walker, N.S., Conte, L.L., Koehl, P., Levitt, M., Brenner, S.E.: The ASTRAL Compendium in 2004. Nucleic Acids Res. 32, D189–D192 (2004)
31. Velankar, S., Best, C., Beuth, B., Boutselakis, C.H., Cobley, N., Sousa Da Silva, A.W., Dimitropoulos, D., Golovin, A., Hirshberg, M., John, M., Krissinel, E.B., Newman, R., Oldfield, T., Pajon, A., Penkett, C.J., Pineda-Castillo, J., Sahni, G., Sen, S., Slowley, R., Suarez-Uruena, A., Swaminathan, J., Van Ginkel, G., Vranken, W.F., Henrick, K., Kleywegt, G.J.: PDBe: Protein Data Bank in Europe. Nucleic Acids Research 38, D308–D317 (2010)

Predicting Rare Classes of Primary Tumors with Over-Sampling Techniques

Nittaya Kerdprasop and Kittisak Kerdprasop

Data Engineering Research Unit, School of Computer Engineering,
Suranaree University of Technology, 111 University Avenue,
Nakhon Ratchasima 30000, Thailand
{nittaya,kerdpras}@sut.ac.th

Abstract. The discovery of hidden biomedical patterns from large clinical databases can uncover knowledge to support prognosis and diagnosis decision makings. Researchers and healthcare professionals have applied data mining technology to obtain descriptive patterns and predictive models from biomedical and healthcare databases. However, clinical application of data mining algorithms has a severe problem of low predictive accuracy that hamper their wide usage in the clinical environment. We thus focus our study on the improvement of predictive accuracy of the models created from the data mining algorithms. Our main research interest concerns the problem of learning a tree-based classifier model from a multiclass data set with low prevalence rate of some minority classes. We apply random over-sampling and synthetic minority over-sampling (SMOTE) techniques to increase the predictive performance of the learned model. In our study, we consider specific kinds of primary tumors occurring at the frequency rate less than one percent as rare classes. From the experimental results, the SMOTE technique gave a high specificity model, whereas the random over-sampling produced a high sensitivity classifier. The precision performance of a tree-based model obtained from the random over-sampling technique is on average much better than the model learned from the original imbalanced data set.

Keywords: Primary tumor classification, Imbalanced data, Over-sampling, SMOTE, Tree-based classifier.

1 Introduction

Human body is made up of many types of cells. Living cells grow and divide to produce new cells in an orderly and controlled manner. However, cell production process can go wrong by continuing to produce new cells even when they are not needed. The result of such event is a mass of extra tissue called a tumor. A primary tumor refers to a tumor that has been developed at the original site where it first generated [19]. A tumor can be benign, which means it does not a cancerous one, or malignant that causes cancer. The cancerous cells can invade nearby tissue or spread to cause secondary tumor in other parts of the body. When the spread occurs, an effective treatment becomes a difficult task. Detecting tumor at its original site is

T.-h. Kim et al. (Eds.): DTA/BSBT 2011, CCIS 258, pp. 151–160, 2011.

therefore important to a successful treatment planning. In this research study, we focus on the problem of detecting and correctly classifying specific types of primary tumors.

In machine learning and data mining, classifying primary tumor data [6] is a difficult task due to the multiclass and imbalanced characteristics inherent in the data set. Many learning algorithms in the past have been proposed to solve the binary classification problem successfully. The problem concerns the finding of a classification model from a given data set to predict either positive, or negative class labels for the new unseen examples. For the data domains with more than two classes, such as text categorization and medical diagnosis, efficient data mining algorithms need some extensions to deal with the multiclass problem. Decision tree induction algorithms [1], [13] use the information theoretic approach to handle data with multiclass, whereas other learning algorithms such as support vector machines employ the serial binary classification techniques including one versus all [5], [14] and some other sophisticated techniques [16], [17], [21]. In this paper, we study the application of decision tree induction algorithm to the multiclass classification problem. Decision tree has the advantage of understandability over other forms of classification models and it has been widely used in the biomedical domain [9], [11], [12].

Another challenging characteristics of primary tumor data is the class distribution imbalance. From the 21 different kinds of primary tumors, some majority classes such as lung and stomach tumors occur three to five times more often than the average frequency rate, while the minority classes occur less than one percent. The primary tumor data set used in our study contains six minority classes, that is, duoden and small intestine, salivary glands, bladder, testis, cervix uteri, and vagina. The minority classes are often missed out by most data mining algorithms because of their extremely low occurrence.

Data mining is about building a model that can best characterize underlying data and accurately predict the class of unlabelled data. The quality of data mining model depends directly on the quality of the training data. Data of low quality are those that contain noise, missing values, and class imbalance. A data set is imbalanced if the number of data instances in one class is much more than those in other classes. In the presence of class imbalance, data mining models are biased toward the majority class in such a way that the models can predict the majority class correctly but data instances from the minority class tend to be incorrectly predicted. This is a problem known as *rare class mining*. This research issue has recently received much attention from the data mining and machine learning research community [3], [4], [15], [18], [20].

In the context of data mining, rare class refers to labeled data instances that are infrequently occurred in the database. Specific problems, such as customer churn prediction [2] and network intrusion detection [8], are more interested in modeling infrequent patterns than the frequent ones. To solve the problem of biased learning toward the majority class, many researchers consider the sampling techniques for manipulating class distribution such that rare class could be sufficiently represented in the training data. The basic sampling techniques that have been applied are under-sampling and over-sampling. Under-sampling alters the class distribution by removing data instances from the minority class, whereas over-sampling duplicates data instances in the minority class [8], [20]. The under-sampling technique may remove good representatives, while over-sampling may cause the over-fitting problem.

In this paper, we apply the two over-sampling techniques to the primary tumor data set, that is, random over-sampling and SMOTE [4]. Overfitting problem has been observed by applying cross validation and holdout methods for classifier performance analysis. We discuss the performance criteria used in this study in Section 2. The experimental design and results are presented in Sections 3. Conclusion appears as the last section of this paper.

2   Performance Analysis Measurement

In data classification, the classifier is evaluated by a confusion matrix. For a binary class problem (positive and negative classes), a matrix is a square of 2×2 as shown in Figure 1. The column represents the outcomes of classifier. The row is a real value of class label. The numbers appeared in each cell of the matrix has different names, that is, TP, FN, FP, and TN. Each acronym can be explained as follows:

TP = number of positive cases that are correctly identified as positive,
FP = number of negative cases that are incorrectly identified as positive cases,
FN = number of positive cases that are misclassified as negative cases, and
TN = number of negative cases that are correctly identified as negative cases.

Predicted class

		Class = +	Class = −
Actual class	Class = +	TP	FN
	Class = −	FP	TN

Fig. 1. A confusion matrix of the binary classification

We assess the model performance based on the five metrics: true positive rate (recall or sensitivity), false positive rate, specificity, precision, and F-measure. The computation methods of these metrics are as follows [3], [10]:

$$\text{TP rate (or Recall, Sensitivity)} = \frac{TP}{TP + FN}$$

$$\text{FP rate} = \frac{FP}{TN + FP}$$

$$\text{Specificity} = \frac{TN}{TN + FP}$$

$$Precision = \frac{TP}{TP + FP}$$

$$F\text{-}measure = \frac{2TP}{2TP + FP + FN}$$

For the case of multiclass classification, a confusion matrix is a square of N×N, where N is the number of classes. The classifier's performance measurement is computed per class. For instances, when N is 3, the confusion matrix can be shown as in Figure 2, and the sensitivity values can be computed as follows:

$$\text{Sensitivity of class A} = \frac{T_A}{T_A + F_{B1} + F_{C1}}$$

$$\text{Sensitivity of class B} = \frac{T_B}{T_B + F_{A2} + F_{C2}}$$

$$\text{Sensitivity of class C} = \frac{T_C}{T_C + F_{A3} + F_{B3}}$$

Predicted class

	Class = A	Class = B	Class = C
Class = A	T_A	F_{B1}	F_{C1}
Class = B	F_{A2}	T_B	F_{C2}
Class = C	F_{A3}	F_{B3}	T_C

Actual class

Fig. 2. A confusion matrix of the three-class classification

3   Experimental Design and Results

The main objective of our experiments is to investigate the advantages of applying random over-sampling and SMOTE [4] techniques to the primary tumor data set. This data set is highly imbalanced in terms of the class distribution (given in Table 1). Our preliminary hypothesis is that by biasing the class distribution of the minority data, the tree-based learning algorithm may perform better on recognizing the rare classes. To make a fair comparison, we use a holdout method, instead of the 10-fold cross validation, to test the classifier performance.

Table 1. Class distribution of the primary tumor data set

Tumor class	Number of cases	Distribution (%)	Rare class
Lung	84	24.8%	
Head and neck	20	5.9%	
Esophagus	9	2.6%	
Thyroid	14	4.1%	
Stomach	39	11.5%	
Duoden and small intestine	1	0.5%	*
Colon	14	4.1%	
Rectum	6	1.7%	
Salivary glands	2	0.5%	*
Pancreas	28	8.2%	
Gallbladder	16	4.7%	
Liver	7	2.1%	
Kidney	24	7.1%	
Bladder	2	0.5%	*
Testis	1	0.5%	*
Prostate	10	2.9%	
Ovary	29	8.5%	
Corpus uteri	6	1.7%	
Cervix uteri	2	0.5%	*
Vagina	1	0.5%	*
Breast	24	7.1%	

The first step of our experimentation is to duplicate a data record containing only a single case (that is, the case of duoden and small intestine, testis, and vagina tumors) to contain two records of each class of tumor. This duplication step is for the purpose of splitting the original data set into two parts: a train set and a test set. Each data set contains the same amount of cases in each type of primary tumors. The independent test data set contains 171 data records. The train data set is to be copied into 3 versions. The first version contains 171 data records with the same class distribution as the test data set. It is called the imbalanced data set. The second version of the train data is to be over-sampling the minority classes with the SMOTE technique [4]. The third version of train data is for the random over-sampling.

We prepare the random over-sampling data set by duplicating data records in each class to be almost the same amount. The maximum number of cases in the majority class is 42, and the minimum number of cases after duplicating is 36. This random over-sampling data set contains 848 data records with the same proportion of class distribution (around 4.2%-4.9%). This data set is thus has a class distribution different from the original data set (the imbalanced data). When we test the accuracy of classifier built from this data set with the 10-fold cross validation method, the true positive rate and precision are extremely high. But these values are much lower when we test the classifier with an independent test set that has different class distribution. This is obviously the overfitting problem. We therefore compare classifiers obtained from different sampling techniques with the holdout method that can better guarantee overfitting avoidance. The comparative results in terms of true positive rate (recall or sensitivity), false positive rate, precision, F-measure, and specificity are given in Figures 3-6. The symbolic codes for different primary tumor types are as follows:

A = salivary glands,   B = bladder,   C = testis,
D = duoden and small intestine, E = vagina,   F = corpus uteri,
G = rectum,   H = cervix uteri,   I = liver,
J = esophagus,   K = prostate,   L = thyroid,
M = colon,   N = gallbladder,   O = head and neck,
P = breast,   Q = kidney,   R = pancreas,
S = ovary,   T = stomach,   U = lung.

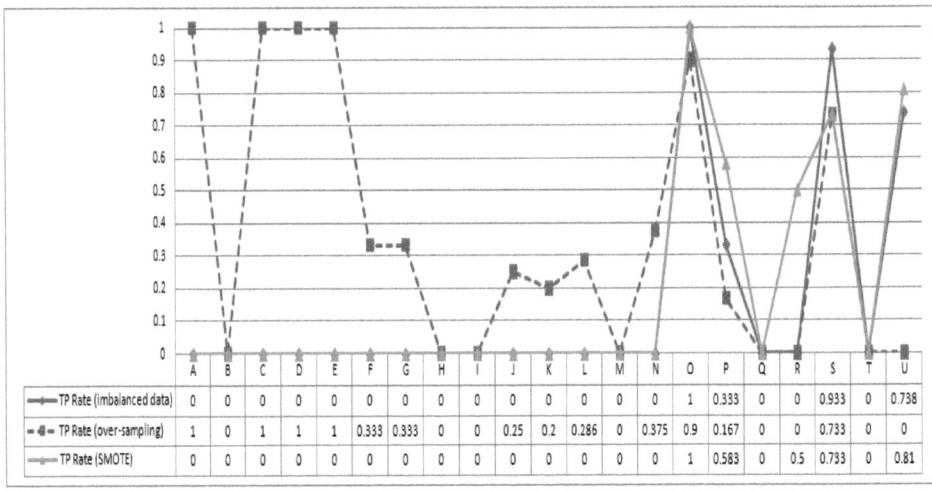

	A	B	C	D	E	F	G	H	I	J	K	L	M	N	O	P	Q	R	S	T	U
TP Rate (imbalanced data)	0	0	0	0	0	0	0	0	0	0	0	0	0	0	1	0.333	0	0	0.933	0	0.738
TP Rate (over-sampling)	1	0	1	1	1	0.333	0.333	0	0	0.25	0.2	0.286	0	0.375	0.9	0.167	0	0	0.733	0	0
TP Rate (SMOTE)	0	0	0	0	0	0	0	0	0	0	0	0	0	0	1	0.583	0	0.5	0.733	0	0.81

Fig. 3. True positive rate (recall or sensitivity) comparison

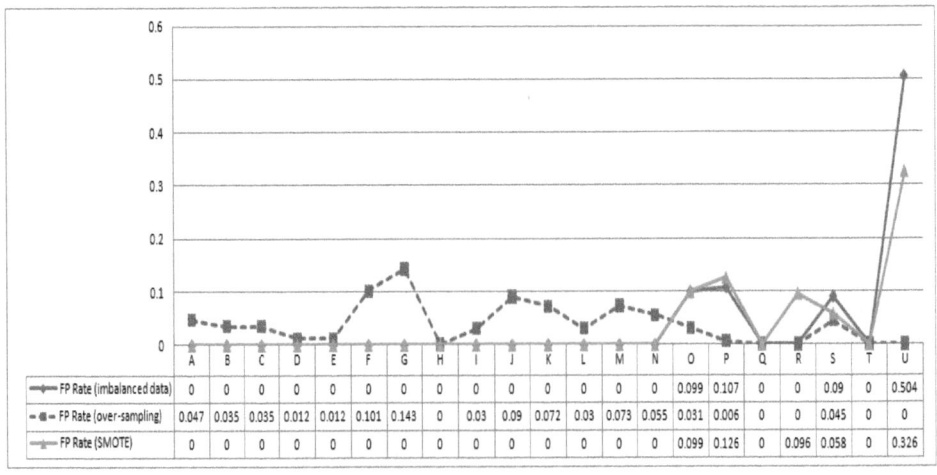

	A	B	C	D	E	F	G	H	I	J	K	L	M	N	O	P	Q	R	S	T	U
FP Rate (imbalanced data)	0	0	0	0	0	0	0	0	0	0	0	0	0	0	0.099	0.107	0	0	0.09	0	0.504
FP Rate (over-sampling)	0.047	0.035	0.035	0.012	0.012	0.101	0.143	0	0.03	0.09	0.072	0.03	0.073	0.055	0.031	0.006	0	0	0.045	0	0
FP Rate (SMOTE)	0	0	0	0	0	0	0	0	0	0	0	0	0	0	0.099	0.126	0	0.096	0.058	0	0.326

Fig. 4. False positive rate comparison

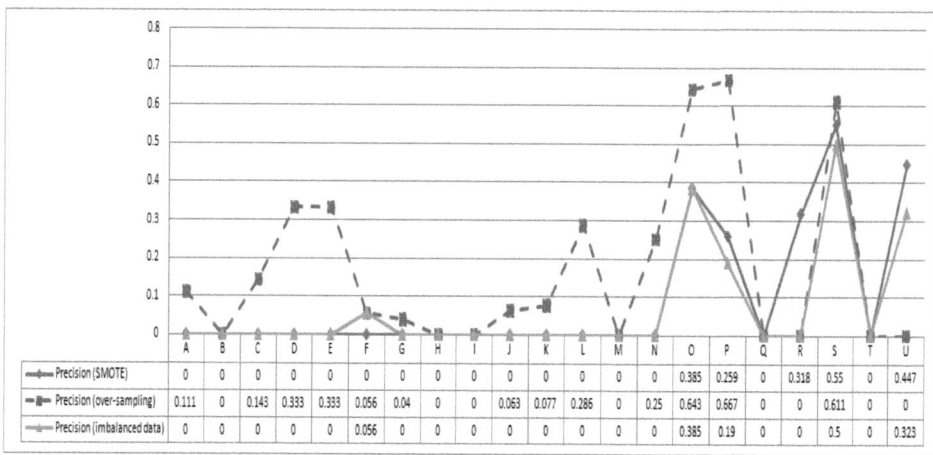

Fig. 5. Precision comparison

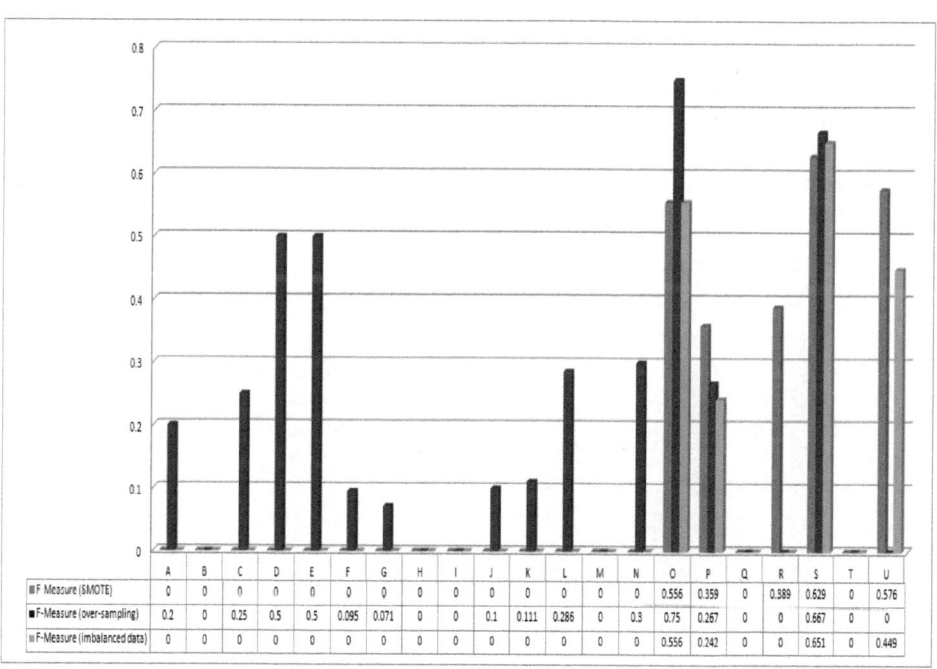

Fig. 6. F-measure comparison.

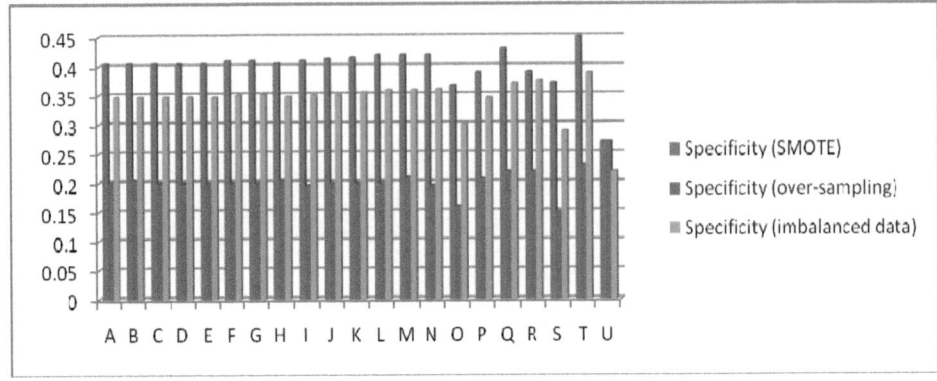

Fig. 7. Specificity comparison

It can be seen from the experimental results that the random over-sampling and SMOTE techniques can improve the performance of predicting rare classes (class A, B, C, D, and E). Random over-sampling yields a better result in terms of sensitivity and precision, whereas the SMOTE technique gives the best sensitivity performance. We also consider the ROC (receiver operating characteristic) area under curve of each technique. The ROC area is a measurement to compare a tradeoff between true positive and false positive error rates. The desired ROC area is over 0.5, and the higher is the better. The ROC area comparison of the two over-sampling techniques against the imbalanced data, specifically for the five most rare classes, is illustrated in Figure 8.

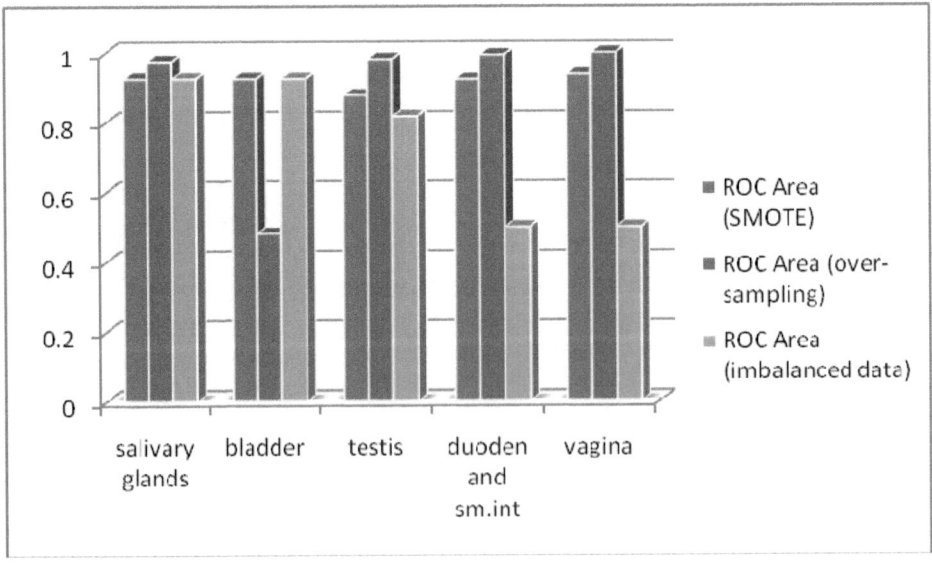

Fig. 8. ROC area comparison of the five rare classes

4 Conclusion

The problem of predicting correctly rarely occurring cases is important to many real life applications including clinical tests and medical diagnosis. Some data domains in the clinical and medical environment are difficult to build an accurate predictive model because of the inherent rare class situation. We thus investigate the over-sampling techniques to bias a decision tree learning algorithm towards the minority classes. In this study, we investigate the random over-sampling and the synthetic minority over-sampling (SMOTE) techniques.

The experimental results show a good predictive performance of both over-sampling techniques on predicting minority classes. Random over-sampling performs slightly better than SMOTE on recalling (or sensitivity test) the minority cases. But in terms of sensitivity (or the true negative rate), SMOTE shows the best performance. From this preliminary results, we plan to conduct a more extensive study on other kinds of medical data domains. The devise of a learning algorithm suitable for such domains is also our future research direction.

Acknowledgments. This work has been supported by grants from the National Research Council of Thailand (NRCT) and Suranaree University of Technology via the funding of Data Engineering Research Unit.

References

1. Breiman, L., Freidman, J., Olshen, R., Stone, C.: Classification and Regression Trees. Wadsworth (1984)
2. Burez, J., Van den Poel, D.: Handling class imbalance in customer churn prediction. Expert Systems with Applications 36, 4626–4636 (2009)
3. Chawla, N.: Data mining for imbalanced datasets: an overview. In: Maimon, O., Rokach, L. (eds.) Data Mining and Knowledge Discovery Handbook, pp. 853–867. Springer, Heidelberg (2005)
4. Chawla, N., Bowyer, K., Hall, L., Kegelmeyer, W.: SMOTE: Synthetic Minority Over-sampling Technique. J. of Artificial Intelligence Research 16, 341–378 (2002)
5. Debnath, R., Takahide, N., Takahashi, H.: A decision based one-against-one method for multi-class support vector machine. Pattern Analysis & Applications 7(2), 164–175 (2004)
6. Frank, A., Asuncion, A.: UCI Machine Learning Repository. University of California, School of Information and Computer Science, Irvine (2010), http://archive.ics.uci.edu/ml
7. Hall, M., Frank, E., Holmes, G., Pfahringer, B., Reutemann, P., Witten, I.H.: The WEKA data mining software: an update. SIGKDD Explorations 11(1), 10–18 (2009)
8. Han, S., Yuan, B., Liu, W.: Rare class mining: progress and prospect. In: Proc. Chinese Conference on Pattern Recognition, pp. 1–5 (2009)
9. Kretschmann, E., Fleischmann, W., Apweiler, R.: Automatic rule generation for protein annotation with the C4.5 data mining algorithm applied on SWISS-PROT. Bioinformatics 17(10), 920–926 (2001)
10. Lalkhen, A.G., McCluskey, A.: Clinical tests: sensitivity and specificity. Continuing Education in Anaesthesia, Critical Care & Pain 8(6), 221–223 (2008)
11. Mugambi, E.M., Hunter, A., Oatley, G., Kennedy, L.: Polynomial-fuzzy decision tree structures for classifying medical data. Knowledge-Based Systems 17(2-4), 81–87 (2004)
12. Pandey, B., Mishra, R.B.: Knowledge and intelligent computing system in medicine. Computers in Biology and Medicine 39, 215–230 (2009)

13. Quinlan, J.R.: Induction of decision tree. Machine Learning 1, 81–106 (1986)
14. Rifkin, R., Klautau, A.: In defense of one-vs-all classification. J. of Machine Learning Research 5, 101–141 (2004)
15. Stefanowski, J., Wilk, S.: Selective pre-processing of imbalanced data for improving classification performance. In: Proc. DaWaK 2008, pp. 283–292 (2008)
16. Tapia, E., Ornella, L., Bulacio, P., Angelone, L.: Multiclass classification of microarray data samples with a reduced number of genes. BMC Bioinformatics 12, 59 (2011)
17. Thabtah, F.A., Cowling, P., Peng, Y.: Multiple labels associative classification. Knowledge and Information Systems 9(1), 109–129 (2006)
18. Van Hulse, J., Khoshgoftaar, T.: Knowledge discovery from imbalanced and noisy data. Data & Knowledge Engineering 68, 1513–1542 (2009)
19. Webster's New WorldTM Medical Dictionary, 3rd edn. Wiley Publishing (2008)
20. Weiss, G.M.: Mining with rarity: a unifying framework. SIGKDD Explorations 6(1), 7–9 (2004)
21. Yeung, K.Y., Bumgarner, R.E.: Multiclass classification of microarray data with repeated measurements: application to cancer. Genome Biology 4(12), R83 (2004)

Mask-Rendering of Mitochondrial Transports Using VTK

Yeonggul Jang[1], Hackjoon Shim[2], and Yoojin Chung[1,*]

[1] Department of Computer Science and Engineering
Hankuk University of Foreign Studies
Kyonggi, 449-791, Republic of Korea
chungyj@hufs.ac.kr
[2] Yonsei University College of Medicine
Cardiovascular Research Institute
50 Yonsei-ro, Seodaemun-gu
Seoul 120-752, Republic of Korea

Abstract. Mitochondria is an important organelle for maintaining cells such as neurons' physiological processes. Mitochondrial transport is known to be strongly related to neurodegenerative disease of the central nervous system such as Alzheimer's disease and Parkinson's disease. Thus, many researchers are interested in mitochondrial transport. Unfortunately, there is no automated tool to analyze the movement and thus correlate them with cellular characteristics. In this paper, we develop a mask-rendering tool and two-dimensional visualization tool based on a slice for analysis of mitochondrial transports using the Visualization Toolkit (VTK).

Keywords: Mask-rendering, mitochondrial transports, VTK.

1 Introduction

Neuro-degenerative diseases of the central nervous system (CNS), e.g., Alzheimer's disease, Parkinson's disease, and multiple sclerosis, have been an intensely researched area due to their fatality and fast growing prevalence [1-3]. It is postulated that these diseases are linked to defective axonal transport in the CNS neurons. Live cell microscopy of the neurons that have been fluorescently labeled with mitochondria have been used to investigate the relationship between the state of axonal transport and health of the neurons. Unfortunately, detailed quantitative analysis of the motions and morphological changes to mitochondria has been difficult due to lack of appropriate image processing methods. This paper describes a new automated image analysis tool that can be used to track and analyze fluorescent images taken by live-cell microscopy.

---

* Corresponding author.
This research was supported by Basic Science Research Program through the National Research Foundation (NRF) funded by the Ministry of Education, Science, and Technology (2009-0069549).

T.-h. Kim et al. (Eds.): DTA/BSBT 2011, CCIS 258, pp. 161–166, 2011.

A novel micro-fluidic cultural platform devised by Taylor *et al.* [4, 7] has accelerated studies on mitochondria transport. Furthermore, it allows for acquisition of time-lapse fluorescent images which display *in vivo* mitochondrial transport. However, analysis using these image sequences is still challenging, because most of those researches [1-3, 4, 7] have been done manually or at most semi-automatically.

In this paper, we develop a mask-rendering tool and two-dimensional visualization tool based on a slice for analysis of mitochondrial transports using the Visualization Toolkit (VTK) [6]. Mask-rendering is a polygon-rendering technique that uses previously segmented images. Two-dimensional (2D) visualization tool based on a slice shows an image divided by x, y and z axes, respectively. We use time-lapse images acquired from neurons growing in a micro-fluidic device [4, 7]. VTK is an open-library package that is specialized for analyzing medical images.

In section 2, we explain the development of a mask-rendering tool and two-dimensional visualization tool based on a slice using VTK. In section 3, we show their experimental result and conclude in section 4.

2   Mask-Rendering and Two-Dimensional Visualization Based on an Image Slice

Volume-rendering aids in visualization in 3D by representing the images and setting opacity and color according to intensity. We visualized time-lapse 2D images of mitochondria movement in 3D that is plotted in time axis. On the other hand, setting opacity not only has function that can see desired intensity wide, but can be used to obtain previously rendered segmentation image.

Mask-rendering is one of polygons-rendering techniques. It visualizes the previously segmented mask image in 3D. It uses volume-rendering or surface-rendering. Mask-rendering using volume-rendering presents the previously segmented mask area into desired color by setting opacity and color. Mask-rendering using surface-rendering extracts the surface of the previously segmented mask area and paints the surface by mapping the color using lookup table.

2D visualization tool based on a slice presents 3D image into 2D slice in x,y and x,z and y,z planes. So, it enables visualization from different angles at specific location.

2.1   Implementation of Mask-Rendering

Mask-Rendering uses volume-rendering or surface-rendering and each consists of two steps: visualization modeling step and graphic modeling step. Works in each step are as follows.

* Visualization modeling step
- vtkImageData : it is image information to be transformed to graphic data.
- vtkContourFilter : it extracts edge from input image data.
- vtkImageMapToColors : it paints input image using a lookup table.

- vtkProbeFilter : it gets two inputs and interpolate one input into the other input.
- vtkPolyDataMapper : it presents an object into a polygon.

* Graphic modeling step
- vtkActor : it represents an object shown in screen in surface-rendering.
- vtkRenderer : it represents a 2D image which is determined by a camera, a light and an actor in 3D space.
- vtkRenderWindow : it manages the rendered 2D image and connects it to a window screen.

This pipeline starts from visualization modeling. MaskImageDataBase is produced from a mask file. MaskImageDataBase is similar to mask data but is initialized as follows to extract edge: If scalar value is above 1, then it is replaced by value 255. If not, it is replaced by value 0. Then, vtkContourFilter extract edges using it. MaskImageData references the lookup table which assigns a RGB value (color) to each intensity and vtkImageMpToColors does color-mapping. Then, MaskImageDataBase gets it with MaskImageDataBase as inputs, paints edges and connects output to vtkPolyDataMapper, which presents an object into a polygon. Then it is represented into 3D object by being connected to Actor. Then, visualization modeling ends and graphic modeling starts.

2.2   Implementation of 2D Visualization Based on an Image Slice

We implement 2D visualization based on a slice by representing 2D image in 3D not in 2D space. Thus this pipeline also starts with visualization modeling and then executes graphic modeling as mask-rendering. Works in each modeling in this pipeline are as follows.

Pipeline of two-dimensional visualization based on a slice same as follows.

* Visualization modeling
- vtkImageData : it is image information to be transformed to graphic data.
- vtkImageViewer2: it gets 3D image as input and shows its selected 2D image.

* Graphic modeling: it is same as that of mask-rendering.

This pipeline also starts from visualization modeling. vtkImageViewer2 gets vtkimagedata as input, set SetSliceOrientationTo function to fit each plane and user set slide control which become an input of SetSlice function. Then GetImageActor function makes vtkImageActor in 3D space from 2D image. Now, visualization modeling finishes and graphic modeling starts. The rest of the pipeline is similar to that of mask-rendering.

3   Experiments

In this section, we describe our experimental results. The size of data in one frame used in our experiment is 476 x 401 and an image consists of 100 frames. Fig. 1. (a) is the result of our mask-rendering and Fig. 1. (b) is the result of combining

mask-rendering with 2D visualization based on a slice. In Fig 2, we can see mitochondrial transports in detail by combining mask-rendering with 2D visualization
based on a slice. In the figure, we can see 2D images where X,Z and Y,Z slices are
combined with time axis, respectively. Fig. 2. (b) shows moving mitochondria of
image. Moving objects are represented as diagonal lines and stopping objects as
straight lines.

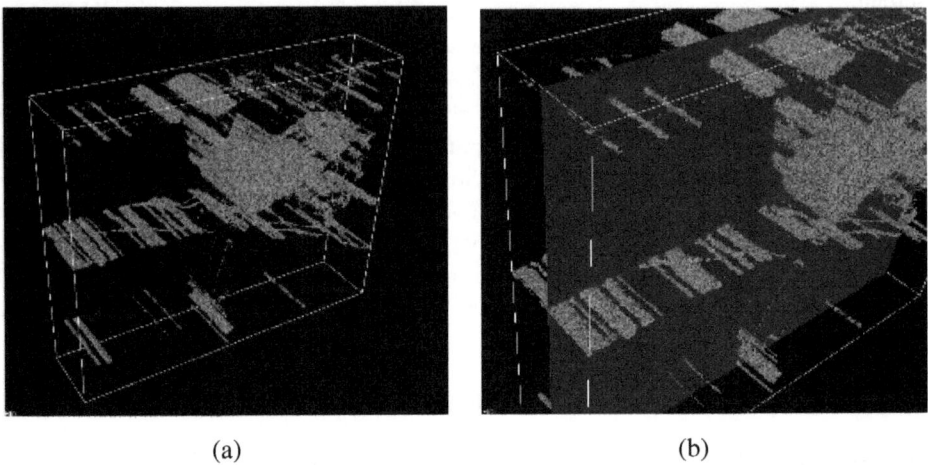

(a)   (b)

Fig. 1. (a) Result of our mask-rendering. (b) Result of combining mask-rendering with 2D
visualization based on a slice.

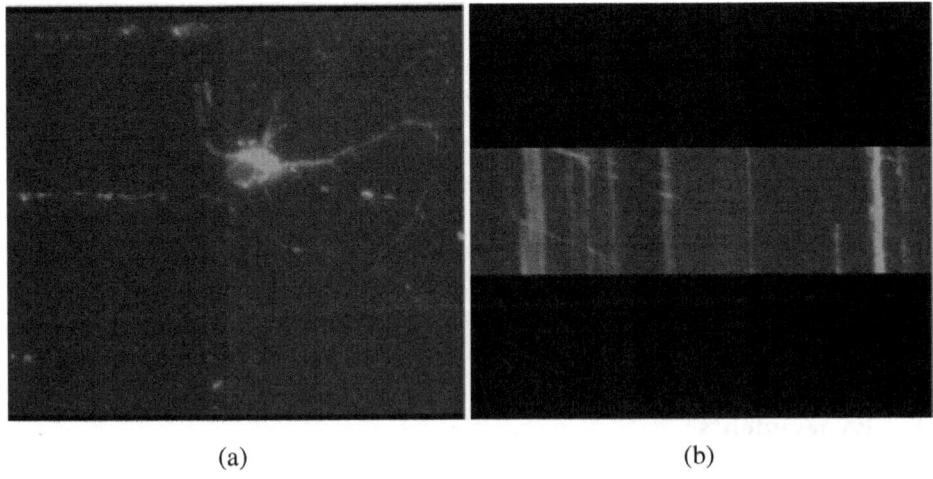

(a)   (b)

Fig. 2. 2D visualization based on a slice : (a) 0 Z Slice. (b) 218 Y Slice. (c) 215 X Slice.

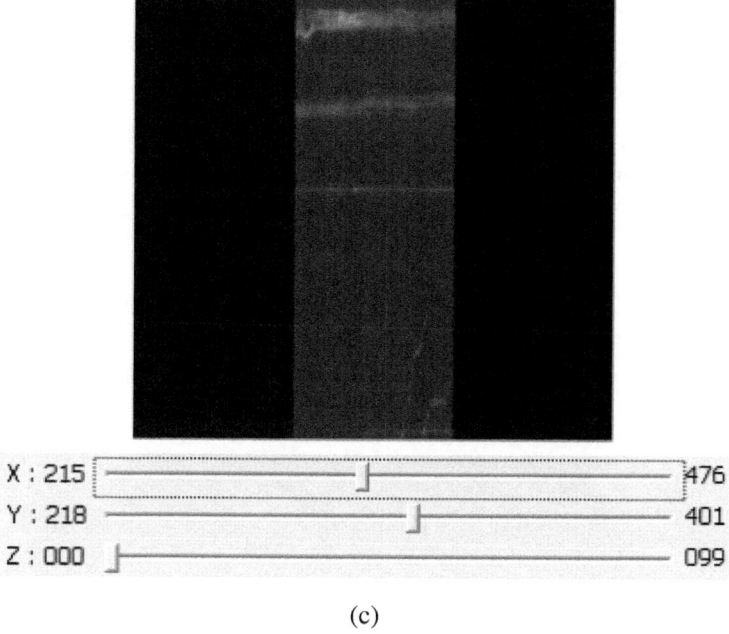

X : 215   476

Y : 218   401

Z : 000   099

(c)

Fig. 2. (*Continued*)

4 Conclusion

In this paper, we develop a mask-rendering tool and two-dimensional visualization tool based on a slice for analysis of mitochondrial transports using VTK. Using these tools, we can observe moving mitochondria in detail. Furthermore, we can observe them in different viewpoint two-dimensional visualization tool based on a slice. And we can see specific part in three-dimensional space in detail by combining mask-rendering with two-dimensional visualization based on a slice.

We delete small connected components to remove noise of image using specific intensity. But, we cannot remove noise completely. In the future, we need to develop efficient noise removing method.

References

1. Malaiyan, L.M., Honick, A.S., Rintoul, G.L., Wang, Q.J., Reynolds, I.J.: Zn2 + inhibits mitochondrial movement in neurons by phosphatidylinositol 3-Kinase activation. Journal of Neuroscience 25(41), 9507–9514 (2005)
2. Cheng, D.T.W., Honick, A.S., Reynolds, I.J.: Mitochondrial trafficking to synapses in cultured primary cortical neurons. Journal of Neuroscience 26, 7035–7045 (2006)
3. Reynolds, I.J., Santos, S.: Rotenone inhibits movement and alters morphology of mitochondria in culture forebrain neurons. Society Neuroscience Abstract 31, 1017–1019 (2005)

4. Taylor, A.M., Blurton-Jones, M., Rhee, S.W., Cribbs, D.H., Cotman, C.W., Jeon, N.L.: A Microfluidic Culture Platform for CNS Axonal Injury, Regeneration and Transport. Nature Methods 2(8) (August 2005)
5. Miller, K.E., Sheetz, M.P.: Axonal mitochondrial transport and potential are correlated. Journal of Cell Science 117, 2791–2804 (2004)
6. Kitware, Inc., The VTK User's Guide: Updated for VTK 4.4, Kitware, Inc. (2004)
7. Park, J.W., Vahidi, B., Taylor, A.M., Rhee, S.W., Jeon, N.L.: Microfluidic culture platform for neuroscience research. Nature Protocols 1(4), 2128–2136 (2006)

Daisyworld in Two Dimensional Small-World Networks

Dharani Punithan, Dong-Kyun Kim, and RI (Bob) McKay

Structural Complexity Laboratory, Seoul National University, South Korea
{punithan.dharani,dkkim1004,rimsnucse}@gmail.com

Abstract. Daisyworld was initially proposed as an abstract model of the self-regulation of planetary ecosystems. The original one-point model has also been extended to one- and two-dimensional worlds. The latter are especially interesting, in that they not only demonstrate the emergence of spatially-stabilised homeostasis but also emphasise dynamics of heterogeneity within a system, in which individual locations in the world experience booms and busts, yet the overall behaviour is stabilised as patches of white and black daisies migrate around the world. We extend the model further, to small-world networks, more realistic for social interaction – and even for some forms of ecological interaction – using the Watts-Strogatz (WS) and Newman-Watts (NW) models. We find that spatially-stabilised homeostasis is able to persist in small-world networks. In the WS model, as the rewiring probabilities increase even far beyond normal small-world limits, there is only a small loss of effectiveness. However as the average number of connections increases in the NW model, we see a gradual breakdown of heterogeneity in patch dynamics, leading to less interesting – more homogenised – worlds.

1   Introduction

Daisyworld was introduced by Watson and Lovelock [1] as an abstract model of planetary self-regulation resulting from the interactions of a multi-species ecosystem. The original model was based on a one-point world, but it was subsequently extended to a one-dimensional (1D) world by Adams et al. [2], to a two-dimensional(2D) world by von Bloh [3] and to a 2D curvature world by Ackland [4].

Although Daisyworld was originally introduced as an ecosystems model, it has seen subsequent application in a number of fields [5, 6], among them abstract simulation of social interactions [7]. Many 2D versions of daisyworld are of regular lattice structure, which may be appropriate for simulations of traditional social networks, governed by short-distance relationships. However modern telecommunications have changed this structure, with long-range links becoming prevalent [8] and small world structures emerging. Even in the original application domain, planetary ecodynamics, species may have a mix of local and global dispersal patterns – for example, coconuts [9] or shellfish [10]. This is the motivation behind the current work, in which we study the effects of global interactions on Daisyworld.

The introduction of a few long range edges into the regular network topology leads to more realistic real-world networks. We study the dynamics of Daisyworld using 2D versions of the Watts-Strogatz (WS) model [11, 12], in which a small fraction of the

T.-h. Kim et al. (Eds.): DTA/BSBT 2011, CCIS 258, pp. 167–178, 2011.

existing edges are rewired as random links, and of the Newman-Watts (NW) model [13], in which a few new random edges are added. These rewired or added edges are known as 'shortcuts', and they give rise to the small-world effect in which quite small numbers of shortcuts can dramatically change many characteristics of the network behaviour. Hence it is of particular interest to determine whether the Daisyworld dynamics are similarly affected.

In this work, we study the change in Daisyworld dynamics as the degree of rewiring or adding increases. In detail, we compare Daisyworld dynamics among completely regular lattice, and WS and NW small-world networks at different degrees of randomisation. For the small-world models, we present results from two different proportions of shortcuts, at the lower and upper end of the ranges generally recognised as exhibiting small-world dynamics.

2   Background

2.1   The Daisyworld Model

Watson and Lovelock [1] initiated the study of planetary ecodynamics with the Daisyworld models, intended as a simple abstract model of the dynamic interaction between planetary biota and environmental driving force. Their zero-dimensional model showed that self-regulation can emerge due to the negative feedback between biotic and abiotic components. This simple self adaptive system has one environmental variable (temperature) and two types of life (black and white daisies). The colour of the daisies influences the environment-altering trait, albedo (i.e. the reflectivity for sunlight), and thereby influences the temperature of the planet. The daisies do not interact with each other directly, only via the temperature of the environment. Black daisies survive better in cooler temperatures, but increase the temperature by absorbing heat; while white daisies survive better at warmer temperatures, but decrease the temperature by reflecting heat. Due to their opposite characteristics, the two species together can self-regulate the temperature through tuning of their population size, without any explicit mechanism for cooperation.

The Logistic Growth Model   Population growth models without density-dependence may lead to geometric or exponential growth in an unlimited environment, which is physically unrealistic. Hence we use the logistic equation with carrying capacity [14] (refer equation 1) which also exhibits complex dynamics in spite of its simplicity.

$$N_{(t+1)} = N_t + rN_t[1 - \frac{N_t}{K}]$$ (1)

where N is population size, t is time step and r is intrinsic capacity for increase.

2.2   Small-World Network Models

Small-world networks [11] are abstract networks characterised by sparseness and rich local clustering, but still possessing short paths between pairs of nodes due to the existence of a few non-local long range connections. Networks having all three properties exhibit the small-world phenomenon known colloquially as 'six degrees of separation'. This degree of separation increases logarithmically with the increasing size of the network. Small-world networks exhibit both localised and distributed structure, thus supporting high dynamic complexity. Small-world characteristics are found both in natural networks, such as the human brain, and in artificial ones such as social networks. They are most commonly based on extending a 1D grid, as in the Watts-Strogatz [11, 12] and Newman-Watts [13] models. Two dimensional small-world models were studied by Kleinberg [15].

The Watts-Strogatz Model: In this model, we start with a pre-existing toroidal grid. The existing edges between the nodes are rewired randomly with probability $p \in [0, 1]$; that is, $pNk/2$ edges are randomly removed, and $pNk/2$ new edges are added, where N is the network size and k is the number of neighbours. When the existing edge from a node is removed, a new edge is added from the same node to a randomly chosen node in the lattice. After the rewiring process, each node has a varying number of edges (degree). The random rewiring does not increase the number of edges, maintaining it as $Nk/2$. Though this WS procedure preserves the number of connections in the original lattice, it does not guarantee that the resulting network is connected.

As p increases, the properties of the network change. When $p = 0$, no existing edge has been rewired, so the network remains a regular network. For smaller values of p, the network remains highly clustered and connected. When $p = 1$, all the existing edges have been rewired, and the network is similar to a random network. In the WS model (refer Figure 1), the connectivity topology can be set anywhere between a completely regular grid and a random network, through tuning the rewiring probability p.

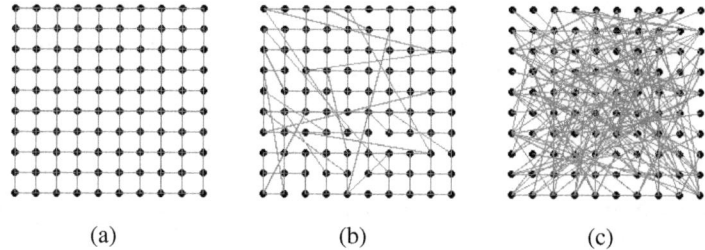

(a)   (b)   (c)

Fig. 1. Watts-Strogatz Model
(a) Regular Network with $p = 0$, $k = 4$ (b) Small-World with $p = 0.05$, $k = 4$
(c) Random Network with $p = 1$, $k = 4$

The Newman-Watts Model: Newman and Watts adapted the WS model to guarantee connectivity, by introducing a variant in which instead of rewiring the links, additional long range links are added with probability $p \in [0, 1]$ between randomly chosen nodes. In contrast with WS, no edges are removed, which means the local connectivity is not altered. After the addition of the extra edges, the number of neighbours at each node can vary but they maintain at least 4 neighbours (if we use the von Neumann (right, left, up and down) neighbourhood). Thus we can guarantee that there are no disjoint nodes.

When $p = 0$, no non local connections are added. When $p = 1$, four extra non local edges are added to each node in the lattice. When p is between 0 and 1, $n \in [0, 4]$ edges are added (refer to Figure 2).

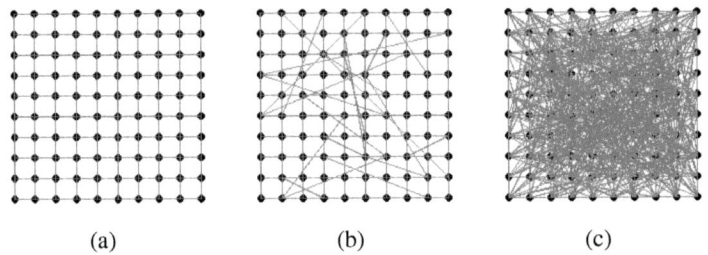

| (a) | (b) | (c) |

Fig. 2. Newmann-Watts Model
(a) Regular Network with $p = 0$, $k = 4$ (b) Small-World with $p = 0.05$, $k = 4$
(c) Network with $p = 1$, $k = 4$

In Figures 1 and 2, we omit periodic boundary conditions for clarity, and also show some connections superimposed.

3   Model

In this section, we describe the overall model we used in our experiments. We first describe the underlying Daisyworld model. It has two parts: the local component describing the effects on growth of daisies and change in temperature local to the particular cell; and the interaction component describing the migration of black and white daisies and diffusion of heat over the links to their neighbours.

Table 1. Daisyworld Parameter Settings

Parameter Name	Value	Parameter Name	Value
Number of cells	100×100	Bare ground Albedo(A_g)	0.5
Heat Capacity(C)	2500	Albedo of black daisies(A_b)	0.25
Diffusion constant(D_T)	500	Albedo of white daisies(A_w)	0.75
Stefan-Boltzmann constant $(\sigma_B)\ E^{-8}Wm^{-2}K^{-4}$	5.67	Mean growth temperature	295.5 K
Luminosity(L)	1	Optimal temp. black daisies	278 K
Solar Insolation(S) Wm^{-2}	864.65	Optimal temp. white daisies	313 K
Initial site population	100	Diffusion rate of black daisies	0.01
Initial temperature at each site	295.5 K	Diffusion rate of white daisies	0.01
Noise Level	5	Natural Rate of Increase	2

3.1   The Underlying Daisyworld Model

Local Dynamics: The population local dynamics is computed using the logistic equation with carrying capacity. The temperature change at each location is based on the energy balance equation [16] (refer equation 2):

$$\sigma_B T^4 = SL(1 - A) \tag{2}$$

where σ_B is the Stefan-Boltzmann constant, T is the temperature, S is the solar constant, L is the luminosity and A is the albedo. The computation of the sub-components is explained below.

Albedo: The albedo of the planet can be computed by equation 3:

$$A = A_b \alpha_b + A_w \alpha_w + A_g \alpha_g \tag{3}$$

where $\alpha_g, \alpha_b, \alpha_w \in [0, 1]$ are the relative areas occupied by bare ground, black and white daisies respectively. α_b and α_w are defined as $\alpha_i = \frac{N_i}{N_{tot}}$, where $N_{tot} = \max(N_b + N_w, K)$ and N_i is the population of daisies of colour i[1]; and we define $\alpha_g = 1 - \alpha_w - \alpha_b$. A_b is the albedo of the ground covered by black daisies, A_w is the albedo of the ground covered by white daisies and A_g is the albedo of the bare ground. We assume, as do most other authors, that $A_w > A_g > A_b$ with corresponding values of 0.75, 0.5, 0.25.

Temperature: The local temperature update consists of terms for the change in temperature due to diffusion, heat radiation and solar absorption, together with a term for Gaussian noise, as in equation 4:

$$C \cdot \delta T_{l,t} = D_T \nabla^2 T_{(l,t)} - \sigma_B T^4_{(l,t)} + SL(1 - A_{(l,t)}) + C\epsilon_{(l,t)} \tag{4}$$

where $C = 2500$ is the heat capacity, $\delta T_{l,t} = T_{l,t+1} - T_{l,t}$ is the change in temperature at (discrete) time t, l indicates a location, $D_T = 500$ is the diffusion constant, $\nabla^2 T$ is the Laplacian operator (generalised to the neighbours in a graph), σ_B is Stefan-Boltzmann constant, S is the solar constant, L is the luminosity, A is the albedo and ϵ represents Gaussian white noise (with mean zero and standard deviation 1.0) multiplied by the noise level.

Growth: The growth rate of daisies is defined as a parabolic function as in equation 5:

$$\beta(T) = 1 - \left[\frac{(T_{opt} - T)^2}{17.5^2}\right] \tag{5}$$

where T is the local temperature, and T_{opt} is the optimal temperature of the species.

---

[1] The long-term average population cannot exceed K, but short-term fluctuations can, so the max formulation is necessary.

Population Size: The local population update consists of terms for the change in population due to Laplacian diffusion, and for local population growth governed by the logistic equation with carrying capacity, as in equation 6:

$$\delta N_{l,t} = D \,\nabla^2 N_{l,t} + r N_{l,t} [\beta(T) - \frac{N_{l,t}}{K}] \tag{6}$$

where at location l and time t, $N_{l,t}$ is the population size, D is the fraction of the population being dispersed to its neighbours, r is the natural rate of increase, $\beta(T)$ is feedback coefficient, and K is the carrying capacity.

We performed preparatory experiments and analysed the system with different settings of the variable parameters, choosing the set of values in Table 1 as a representative set leading to spatially dynamic coexistence on a simple 2D toroidal lattice.

3.2   The Network Structure

All experiments were run on a small-world network generated from a toroidal 2D complete square regular lattice (an $n \times n$ square grid with edges joined both horizontally and vertically) using a size 4 von Neumann (right, left, up and down cells) neighbourhood. We introduced small-world effects in the lattice using either the WS or NW model.

In the WS model, for each edge $u - v$ in the underlying square lattice with k neighbours, the selected edge is rewired with probability p. In other words, a new edge $u - w$ is added to a randomly chosen node $w \notin (u \cup nbhd(u))$ in the rest of the lattice, and the existing edge $u - v$ is removed. In the NW model, for each edge $u - v$ in the underlying square lattice with k nearest neighbours, an extra short cut $u - w$ is added, joining u to a randomly chosen node $w \notin (u \cup nbhd(u))$ with probability p. In both the models, edges are made randomly and independently of the spatial locations; the difference between WS and NW models is that in the latter, the original edges are not deleted. The added or rewired edges are bidirectional (i.e. the graphs are undirected graphs) and the process ensures that self connections (loops), duplicated edges and re-creation of previously removed edges do not occur.

Each cell is viewed as a habitat with a maximum carrying capacity of 10,000 individuals. Each is initialised with the same temperature and population size. The black and white daisies diffuse over their neighbourhood linkages (whether local or long-range). In real-world situations, the long range diffusion of daisies might be due to wind, water or other means. The resource (temperature) also diffuses over all connections – in future work, we will investigate the effect of differing connectivity topologies for resource and species diffusion, but for simplicity, in this report, we only consider common connectivity topologies. We use a Laplacian diffusion model for both species and temperature. Thus the interaction pattern is determined by the network topology and the diffusion term.

Rewiring Probabilities: In the experiments, we used for the rewiring probabilities $p \in \{0.0, 0.05, 0.3, 1.0\}$, giving at the one extreme, completely regular lattices, and at the other, near-random networks (especially for WS). The intermediate values were chosen as representative small-world networks – 0.05 as a generally-accepted value for the onset of small-world effects [17, 18], and 0.3 as a value which is still far from a random network, but showing some of the further changes toward random networks.

4 Results

We use visual inspection of images of the system to determine the dynamics. Here, we show representative frames from the evolution of Daisyworld in order to characterise the dynamics. Specifically, in all cases we show the state of the Daisyworld at time step 1000 (i.e. after the system's behaviour has stabilised from any initial behaviours), 2000 (so that the system's general behaviour can be compared over a long period), 2010 (so that we can see any short-term stability) and 5000 (the limit of the runs). In these visualisations, a location is shown as black if there are only black daisies present, and correspondingly for white. If the species are mixed, then it is shown as dark grey if the total population is over half the carrying capacity (i.e. $N_b + N_w > \frac{K}{2}$), otherwise (including the case where no daisies are present) it is shown as light grey.

4.1 Regular Networks

Figure 3 shows the behaviour of the chosen parameter settings in a regular 2D lattice. We see overall a rich patterning, with regions consisting solely of black or of white daisies, and other regions of coexistence. This structured patterning persists over thousands of epochs, yet the individual patches change on a relatively short time scale, as we can see by comparing sub-figures (b) and (c) of Figure 3, showing the structure only ten epochs apart. Our aim in the rest of these experiments is to see whether these dynamics are maintained in small-world versions of these structures.

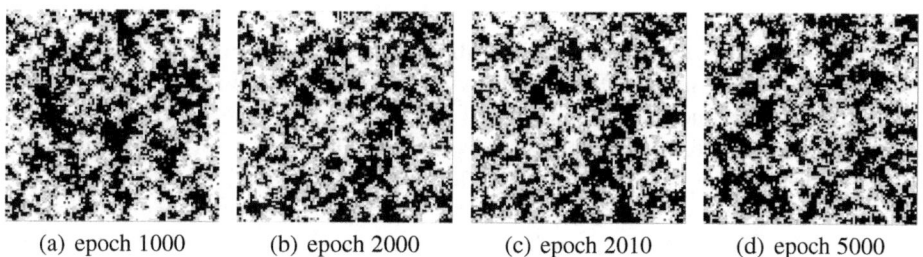

(a) epoch 1000   (b) epoch 2000   (c) epoch 2010   (d) epoch 5000

Fig. 3. Population Structure, Regular 2D 100 × 100 Network

4.2 Small-World Models

Small World with Low Rewiring Probabilities (p=0.05): It is generally accepted in the field [17, 18] that the small-world regime covers at least the region $p \in (0.01, 0.1)$, hence we use $p = 0.05$ as a generally-acknowledged small-world environment. A qualitative assessment of Figures 4 (WS) and 5 (NW) indicates that the behaviour is little different from that seen in the regular lattice – that is, the small-world effect *does not*

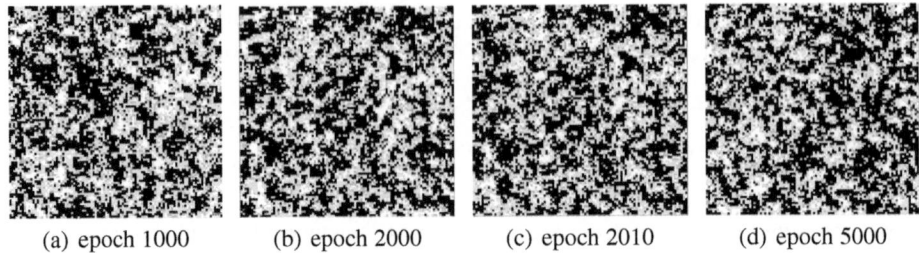

| (a) epoch 1000 | (b) epoch 2000 | (c) epoch 2010 | (d) epoch 5000 |

Fig. 4. Population Structure, Watts-Strogatz 100×100 Network, with $p = 0.05$

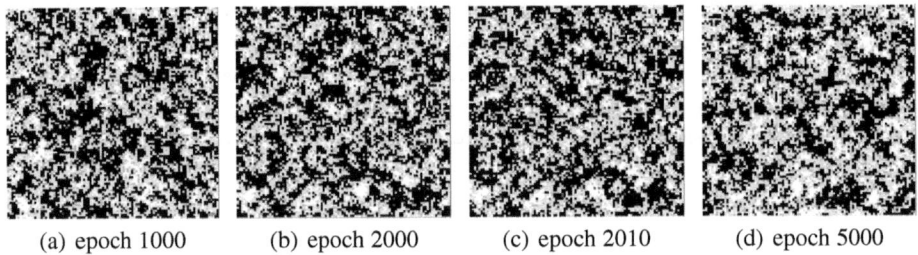

| (a) epoch 1000 | (b) epoch 2000 | (c) epoch 2010 | (d) epoch 5000 |

Fig. 5. Population Structure, Newman-Watts 100×100 Network, with $p = 0.05$

affect the ability of Daisyworld to spatially self-stabilise. This may seem a little surprising in view of the often very large effects that small-world topology may have on system behaviour, but it seems that, in the case of Daisyworld dynamics, such effects are not detectable.

Small World with High Rewiring Probabilities (p=0.3): To confirm whether small-world topology had much effect on Daisyworld dynamics, we ran further tests with $p = 0.3$, well beyond the accepted small-world boundary. The results are shown in Figures 6 and 7. At first sight, the results seem very different, with patches much less visible. However there is a confounding effect in play. The small-world rewiring might be affecting the Daisyworld dynamics – or it might merely be affecting our ability to perceive them. By the time we reach a rewiring probability of $p = 0.3$, the neighbours that influence a cell may no longer be those visually closest to it, and this effect will amplify as distance increases.

To give some perspective on this issue, we took the Daisyworld of Figure 3, and re-plotted it, using the same data, but mapping every location to a random new location. The result is shown in Figure 8. We emphasise that this figure has exactly the same dynamics as Figure 3. All that has changed is the spatial relationship between locations, and hence our inability to perceive patches. The moving patches are still present in the data, but they are impossible to perceive because we no longer have the cue of spatial relationships. Similar effects will undoubtedly be present, though at lower levels, in visualisations of small-world networks by the time we have reached rewiring probabilities of $p = 0.3$.

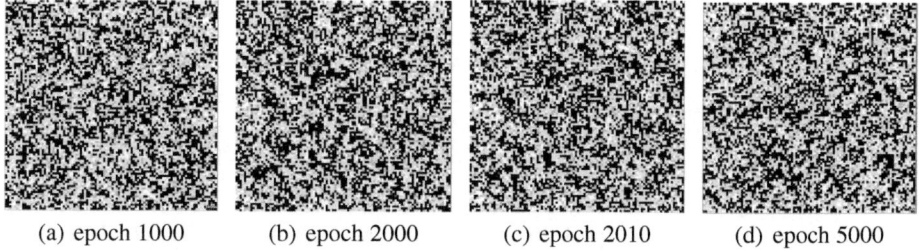

| (a) epoch 1000 | (b) epoch 2000 | (c) epoch 2010 | (d) epoch 5000 |

Fig. 6. Population Structure, Watts-Strogatz 100×100 Network, with $p = 0.3$

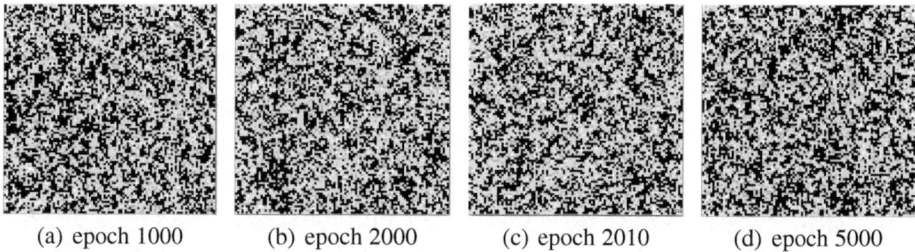

| (a) epoch 1000 | (b) epoch 2000 | (c) epoch 2010 | (d) epoch 5000 |

Fig. 7. Population Structure, Newman-Watts 100×100 Network, with $p = 0.3$

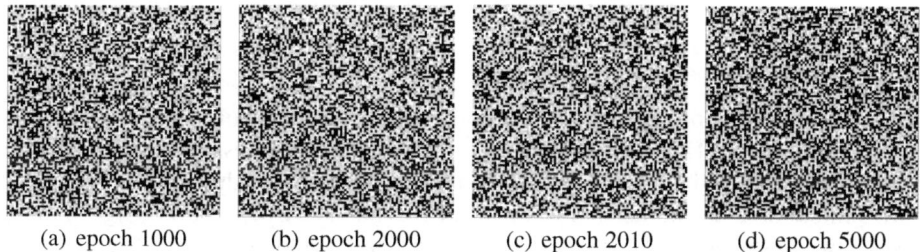

| (a) epoch 1000 | (b) epoch 2000 | (c) epoch 2010 | (d) epoch 5000 |

Fig. 8. Population Structure, Regular 2D 100×100 Network, Randomised

It is important to note that this visualisation effect will be greater in WS than in NW small-world networks, since in the former, 30% of the links are random, while in the latter, only 23% are.

Taking these issues into account, it seems clear that even at a rewiring probability of $p = 0.3$, the Daisyworld dynamics have been only slightly affected, and we are justified in claiming that Daisyworld dynamics are retained in small-world networks. To further investigate this situation, we extended the analysis to an extreme rewiring probability, $p = 1.0$.

4.3 Small World Models Extended towards Random Case (p=1.0)

Figures 9 (WS) and 10 (NW) depict the spatial dynamics for $p = 1.0$. For WS, it is difficult to see any difference from Figure 8, suggesting that even in this extreme small-world case, the Daisyworld spatial dynamics are maintained. On the other hand, Figure 10 does appear to show an enhanced proportion of grey cells: the world is substantially more homogenised; the heterogeneity in patch dynamics begins to disappear.

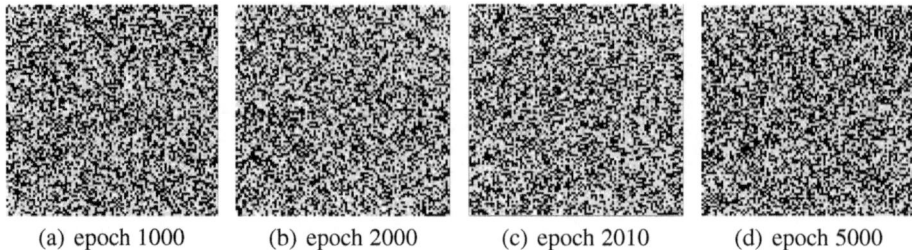

| (a) epoch 1000 | (b) epoch 2000 | (c) epoch 2010 | (d) epoch 5000 |

Fig. 9. Population Structure, Watts-Strogatz 100×100 Network, with $p = 1.0$

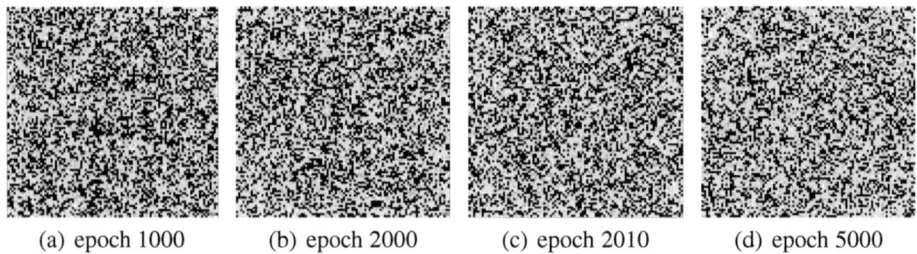

| (a) epoch 1000 | (b) epoch 2000 | (c) epoch 2010 | (d) epoch 5000 |

Fig. 10. Population Structure, Newman-Watts 100×100 Network, with $p = 1.0$

5   Discussion

The most important point to note in these results is the relatively small influence of small-world effects. That is, at low values of p, when nevertheless small-world effects are well-established, there is very little effect on the behaviour of the Daisyworld system – we still see spatial stabilisation of the system. This is, perhaps, surprising. In many other network properties, the change from grid to small-world dynamics is dramatic [19, 20, 21]. It is not particularly obvious why Daisyworld dynamics are so different.

On the other hand, increasing connectivity has a greater effect. Thus we see more substantial change in the NW model than in the WS model. As the number of added links increases($p = 1.0$), we have a much more homogenised behaviour.

This, of course, has implications in many Daisyworld application domains. In ecosystems dynamics, it suggests that it is not the length of connections that is important, but rather the number. Worlds with many long-distance ecological connections may still retain interesting dynamics, so long as the links remain sparse. It is increasing numbers of links, rather than increases in the average length of the links, that leads to a homogenised world.

6   Conclusions

6.1   Summary

The overall conclusions are relatively easy to summarise: in Daisyworld models, long-range connections, such as are important in small-world models, have only limited effects on Daisyworld dynamics, and spatially stabilised dynamics can still be readily observed. On the other hand, increasing link density appears to have a substantial effect, leading – at high link densities – to a relatively homogenised world with reduced dynamicity.

6.2 Assumptions and Limitations

In general, the underlying assumptions of this work are the same as those for all Daisyworld models – namely that Daisyworld provides a realistic and useful analogue for interesting ecosystems and other real-world processes. However this work has revealed limitations in the subjective visualisations used to describe the behaviour of distributed Daisyworlds. We are actively developing objective metrics for assessing the degree of spatial stabilisation in Daisyworld models, and hope to be able to report on this in the relatively near future.

6.3 Future Work

In the immediate future, in addition to developing more objective metrics of Daisyworld behaviour, we are aiming to extend our analysis of the effects of differing network topologies on Daisyworld behaviour. To date, all work in this area, both ours and others, has assumed a common connectivity topology for resources and species. It is possible that further new behaviours might emerge if these were separated – if species had different connectivities, or if their connections were different from those of resources. Again, this is a planned direction of future research.

Acknowledgements. This research was supported by Basic Science Research Program through the National Research Foundation of Korea(NRF) funded by the Ministry of Education, Science and Technology(Project No. 2011-0004338), and the BK21-IT program of MEST. The ICT at Seoul National University provided research facilities for this study.

References

1. Watson, A.J., Lovelock, J.E.: Biological homeostasis of the global environment: the parable of daisyworld. Tellus B 35(4), 284–289 (1983)
2. Adams, B., Carr, J., Lenton, T.M., White, A.: One-dimensional daisyworld: Spatial interactions and pattern formation. Journal of Theoretical Biology 223(4), 505–513 (2003)
3. Von Bloh, W., Block, A., Schellnhuber, H.J.: Self-stabilization of the biosphere under global change: a tutorial geophysiological approach. Tellus B 49(3), 249–262 (1997)
4. Ackland, G., Clark, M., Lenton, T.: Catastrophic desert formation in daisyworld. Journal of Theoretical Biology 223(1), 39–44 (2003)
5. Saunders, P.T., Koeslag, J.H., Wessels, J.A.: Integral rein control in physiology ii: a general model. Journal of Theoretical Biology 206(2), 211–220 (2000)
6. Dyke, J.G., Harvey, I.R.: Pushing up the daisies. In: Artificial Life X, Proceedings of the Tenth International Conference on the Simulation and Synthesis of Living Systems, pp. 426–431. MIT Press (2006)
7. Nuño, J.C., De Vicente, J., Olarrea, J., López, P., Lahoz-Beltrá, R.: Evolutionary daisyworld models: A new approach to studying complex adaptive systems. Ecological Informatics 5(4), 231–240 (2010)

8. Leskovec, J., Horvitz, E.: Planetary-scale views on a large instant-messaging network. In: Proceeding of the 17th International Conference on World Wide Web, pp. 915–924. Association for Computing Machinery (2008)
9. Perera, L., Russell, J.R., Provan, J., Powell, W.: Use of microsatellite dna markers to investigate the level of genetic diversity and population genetic structure of coconut (Cocos nucifera L.). Genome 43(1), 15–21 (2000)
10. White, C., Selkoe, K.A., Watson, J., Siegel, D.A., Zacherl, D.C., Toonen, R.J.: Ocean currents help explain population genetic structure. Proceedings of the Royal Society B: Biological Sciences 277(1688), 1685 (2010)
11. Watts, D.J., Strogatz, S.H.: Collective dynamics of small-world networks. Nature 393(6684), 440–442 (1998)
12. Watts, D.J.: Small Worlds: the Dynamics of Networks between Order and Randomness. Princeton University Press, Princeton (2003)
13. Newman, M., Watts, D.: Renormalization group analysis of the small-world network model. Physics Letters A 263(4-6), 341–346 (1999)
14. Krebs, J.R., Davies, N.B.: An Introduction to Behavioural Ecology. Wiley-Blackwell (1993)
15. Kleinberg, J.M.: Navigation in a small world. Nature 406(6798), 845 (2000)
16. McGuffie, K., Henderson-Sellers, A.: A Climate Modelling Primer, vol. 1. Wiley, Chichester (2005)
17. Kanamaru, T., Aihara, K.: Roles of inhibitory neurons in rewiring-induced synchronization in pulse-coupled neural networks. Neural Computation 22(5), 1383–1398 (2010)
18. Kitano, K., Fukai, T.: Variability vs synchronicity of neuronal activity in local cortical network models with different wiring topologies. Journal of Computational Neuroscience 23(2), 237–250 (2007)
19. Moore, C., Newman, M.E.J.: Epidemics and percolation in small-world networks. Physical Reviews E 61(5), 5678–5682 (2000)
20. Lago-Fernández, L., Huerta, R., Corbacho, F., Sigüenza, J.: Fast response and temporal coherent oscillations in small-world networks. Physical Review Letters 84(12), 2758–2761 (2000)
21. Latora, V., Marchiori, M.: Efficient behavior of small-world networks. Physical Review Letters 87(19), 198701 (2001)

A Verification Tool for Splice Junction Sites on Whole Genome with Massive Reads

Sora Kim[1], Taewon Park[1], KieJung Park[2], and Hwan-Gue Cho[1]

[1] Department of Engineering, Pusan National University, Korea
[2] Division of Bio-Medical Informatics, National Institute of Health, Korea
{srkim_11,darkptw,hgcho}@pusan.ac.kr,
kjpark63@gmail.com

Abstract. Traditionally, methods for detecting junctions considerably rely on the annotation of gene structures. However, ab initio junction detection tools have now become available. Using RNA-Seq data, these tools will find the splice junction sites without annotation of gene structures. The optimum results of using these tools depend on their algorithms. The way to judge the best result is dependent on their algorithms too. In addition, it is hard to verify the results and other candidate splice junction sites independently. In this paper, we introduce our tool, VETic, which can choose results from tools for detecting splice junctions and then search other candidate splice junction sites. This method helps to find other candidate splice junction sites and calculate the reliability of results.

Keywords: NGS, verification, splice junction, alignment.

1 Introduction

Genome sequencing means decoding DNA sequencing information. Genome sequencing is the key to identify individual differences and national characteristics, or to find genes and genetic defects. In addition, sequencing data is important for widespread use in the area of molecular diagnostics and therapy by gene expression, genetic diversity and its interactions with other information. With the development of NGS (Next Generation Sequencing) technology, massive amounts of sequence information can be analyzed at low-cost and more easily compared to traditional methods, but its limitation includes the accurate determination of full DNA sequence information. However, in many fields including transcriptomicse, NGS technology is expected to replace EST (Expressed Sequence Tags) or microarray technology.

Most genome sequencing information using NGS technology has adopted paired-end approaches. It is predicted that many thousands of SNP, InDel, structural variation, CNV. Comparative analysis of DB related to variants information, can be used for disease prediction and diagnosis. Although environmental factors cause disease, most are associated with DNA changes in the cell. After completion of the HGP, it was able to analyze the relevance with disease and identify differences in genomic mechanisms.

T.-h. Kim et al. (Eds.): DTA/BSBT 2011, CCIS 258, pp. 179–186, 2011.
© Springer-Verlag Berlin Heidelberg 2011

Microarray technology in gene expression studies has many limitations. One disadvantage is that the probe is designed by known genes. Also, a microarray is not accurate when expression by detector noise is low or high. There tends to be a large deviation between experiments. In addition, an exon microarray is relatively more expensive than gene expression microarray. This requires normalization of data, including the need to apply appropriate statistical methods. Many researchers find it difficult to apply appropriate statistical methods, including the normalization of data. RNA-Seq is alternative technology that looks at gene and exon expression and variation in expressed genes simultaneously.

RNA-Seq has emerged as a new technology instead of ESTs and microarray thanks to the development of next generation sequencing technology[1][11]. RNA-Seq can represent the gene expression level better than microarray technology. RNA-Seq has advantages in that genes of unknown species may also be applied. Many novel alternative splicing are found by experiments using RNA-Seq. Alternative splicing is a phenomenon in which proteins are created by a variety of different combinations of exons in a given gene. Alternative splicing occurs as a universal phenomenon in eukaryotes and is a ubiquitous post transcriptional process. Early studies of alternative splicing were mainly based on EST libraries[1]. Alternative splicing is considered to occur in 74% of all human genes.

However, recently, MIT Burge of United States assures that high throughput sequencing data of six tissue including, brain, liver and lungs reveals 92-94% of the genes produce two or more mRNA[16]. Blencowes team from the University of Toronto, Canada showed that 94% of genes in the human genome create more than one mRNA in 15 different tissues[3]. Higher organisms such as humans make a wide variety of proteins by increasing the utilization of genetic strategies, rather than evolutionarily increasing the number of genes. Alternative splicing occurs much more than we expected and this produces multiple mRNA transcripts from a single gene.

We first have to find a splice junction to find alternative splicing. Traditionally, the standard method for detecting splice junctions used expressed sequence tags (ESTs) or Sanger sequencing technology. Recently, tools for *ab initio* junction detection such as TopHat[14], SpliceMap[2], MapSplice[15], HMMsplicer[4], SOAPsplice[5] were developed. These tools find the splice junction site using RNA-Seq data. The optimum results of using these tools depend on their algorithms. It is hard to verify the results and other candidate splice junction sites independently. Therefore, we developed a tool for genome analysis and verification of splice junction site. Using this tool, we can reconsider splice junction site on reference sequence from tools for detecting splice junction. Also, this splice junction site appears in the reference sequence that shows information on how many times it can be seen at once. Therefore, our tool can calculate reliability for splice junction sites.

2   Methods

Algorithm of tools for detecting splice junction is divided into three steps.

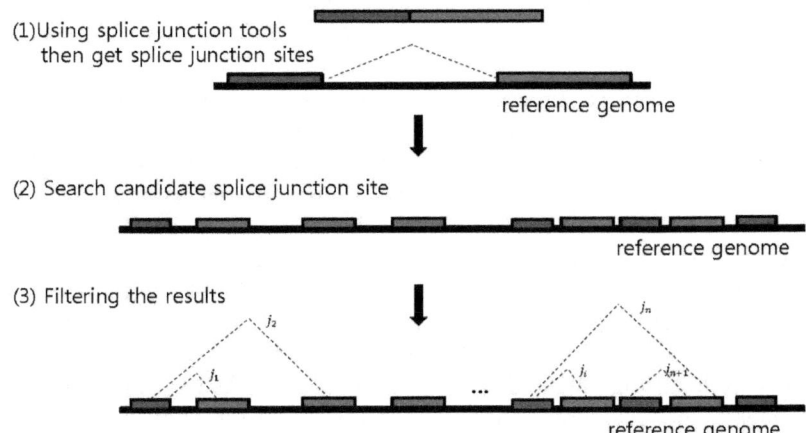

Fig. 1. An overview of our tool pipeline. In step 1, Get information of splice junction sites using tools for detecting splice junction. Then step 2, Align each segment of splice junction. Finally step 3, Find candidate splice junction sites. j is a candidate splice junction site.

First, tools for detecting splice junction separate reads with splice junction and reads without splice junction using their alignment algorithm or another alignment tool. At this step, tools for detecting splice junction will find reads that are expected to have splice junction sites among the hundreds of thousands of reads. Second, it finds actual splice junction site using the reads. At this point, the algorithm can find a splice junction depending on the tools. TopHat finds putative exon using Bowtie[8], and then IUM (Initially UnMapped) reads remap to putative exon. SpliceMap cuts a read in half, and then maps one side of read after the other using Bowtie or SeqMap[6]. MapSplice finds splice junction site using a finely cut tag from the read. HMMSplicer maps the reads to the reference sequence through machine learning using the Hidden Markov Model. The HMM is trained on a subset of read-half alignments to best reflect the quality and base composition of the dataset and genome. Finally to analyze the most recently released SOAPsplice, SOAPsplice divides the IUM reads into two segments, which are expected to be derived from different exons in the premature mRNA. Then, each segment is mapped to the reference. The third step is result filtering. The goal of this step is to create high accuracy by filtering the results from the second phase.

We have developed a system to measure the objective reliability of results obtained from these tools. Reusing splice junction sites from splice junction detecting tools, VETic can verify how many sites exist in the actual reference sequence. Using this method, the splice junction site can be calculated to some degree of reliability.

Our program, as shown in Figure 1 consists of the following three steps.

First, it looks for splice junction sites using existing tools for detecting splice junction. At this time, most tools produce bed format file[13]. Bed format file is descriptive and helps identify where the splice junction site is from the reference sequence. On the other hand, Bed format file present splice junction site position from reference sequence.

Second, VETic finds candidate splice junction sites by using the splice junction site found by the preceding step. When a splice junction site appears elsewhere in reference sequence, it can be considered a candidate splice junction site. The algorithm allows only 2 mismatches at this stage. This value, depending on the user's selection to a value between 0-2, can be changed.

Third, as shown in Figure 1, if there is segment A and segment B, our algorithm can create candidate splice junction sites listed in the gap as long as the maximum length is based on segment A. For instance, segment A is mapped to somewhere specific location, and then if segment B is mapped to three somewhere specific locations in the 50,000bp from segment As location. We associate each segment with segment A and we visualize a group that consists of three combinations, the segment A and a segment mapped with segment B. Using this approach, users can verify splice junction site results that tools for detecting splice junction have ignored or not found previously. The default value of the gap is set to 50,000 bp, because distance of two segments, which is equal to the size of an intron, is expected to range from 50 to 50,000bp, since this range covers the majority of known intron sizes in eukaryotes[14].

The above steps are able to detect just one junction. Reliability can be calculated to verify the splice junction in other sectors of the reference sequence, in other words, by verifying the distribution of the different sectors of the splice junction site. The higher the false positive and false negative rates are, the more similar the value will be. In addition, if a splice junction site was matched to true set, we could expect the average of reliability closing to 1 made by result of VETic using a proper tool, since the true set had to be unique.

3   Results

Figure 2 is a screen shot of the program. In the current version, four of the five tools described above are designed to be used. We will update the program, VETic, after modification that parsed for SOAPsplice output format.

We used dataset of SOAPsplice to verify reliability. They made four sets of simulated reads (with read length = 50, 75, 100, and 150 bp respectively). They also made five type sequencing depths using the short-read simulator from MAQ. However, we only used 50-fold sequencing depth. They obtained data from human chromosome 10. A total of their data is 1,296 RefSeq[12] transcripts. These transcripts contain 8,266 junctions. Table 1, 2, and 3 show our experiment result when VETic takes different number of mismatches. Experiments were progressed for three tools due to dataset of paired-end reads. HMMsplicer is only run with single end reads, and therefore could not be used.

The size of the gap during the experiment was set from 50 to 50,000 bp. This limit covers most of the known intron sizes in eukaryotes[14][5]. The number of multi-mapping was limited to 30,000. After setting, VETic ran with 8,374 to the

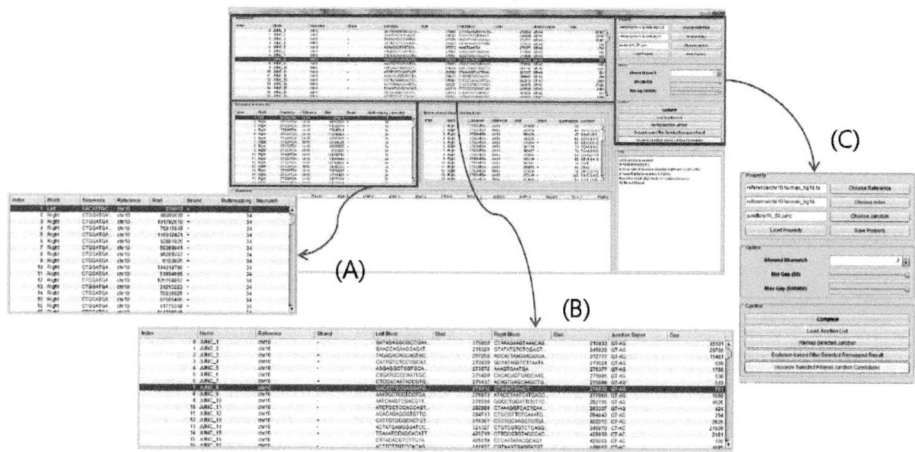

Fig. 2. A snapshot of VETic. VETic is made for windows OS. (A) is list of candidate splice junction sites. (B) is list of splice junction sites from tools for detecting splice junction. (C) is option of run-time configuration.

Table 1. VETic results used simulated data from SOAPsplice. The maximum allowed number of mismatches is 0. # splice junction site means number of splice junction site. C is candidate splice junction site.

	TopHat	SpliceMap	MapSplice
# splice junction site	8374	8451	9723
average of C	1.101	1.046	1.406
minimum of C	1	1	1
maximum of C	12	6	190
# error	248	267	1078

Table 2. VETic results used simulated data from SOAPsplice. The maximum allowed number of mismatches is 1.

	TopHat	SpliceMap	MapSplice
# splice junction site	8374	8451	9723
average of C	1.168	1.076	6.841
minimum of C	1	1	1
maximum of C	33	54	5808
# error	283	262	964

splice junction site from TopHat. After that, the minimum number of candidate splice junction sites was 1, maximum number of candidate splice junction site was 12, and the average number of candidate splice junction sites from the total 8,374 splice junction sites was 1.101. Most of the splice junction sites did not have candidate splice junction sites when mismatch value was 0. 248 mean total

Table 3. VETic results used simulated data from SOAPsplice. The maximum allowed number of mismatches is 2.

	TopHat	SpliceMap	MapSplice
# splice junction site	8374	8451	9723
average of C	1.224	1.080	23.171
minimum of C	1	1	1
maximum of C	48	54	11598
# error	241	262	1017

number of error. An error is detected when the splice junction sites are more than 30,000 multi-mappings to the reference sequence.

SpliceMap was created approximately one year later than TopHat. SpliceMap is smaller than TopHat for the average number of candidate splice junction sites and maximum candidate splice junction sites. In the TopHat case, results rely on reads, because TopHat finds putative exons using Bowtie. Then TopHat finds splice junction site on the putative exons. Therefore, the total number of splice junction sites is smaller than for SpliceMap and MapSplice. In SpliceMap, the average number of candidate splice junction sites is close to 1. In other words, SpliceMap has higher reliability than the other tools.

However, the average number of candidate splice junction sites is 1.406 in MapSplice when maximum allowed number of mismatches is 0. MapSplices average number of candidate splice junction sites is more than TopHat. These results are worse when the number of mismatches is increased. When mismatch value is set to 2, in TopHat and SpliceMap, average number of candidate splice junction sites is similar to mismatch value set to 0. However, in MapSplice, that value is very elevated. The maximum number of candidate splice junction site also represents the 11,598 high values. MapSplice has higher false positive rates compared to other tools. For this reason, we could show that using VETic for reliability by MapSplice results was higher than the other tools using VETic.

These results show that, our system produces better candidate splice junction sites than other tools. This implies that the average number of candidate splice junction sites is close to 1. In addition, VETic produces more maximum number of candidate splice junction sites and average number of candidate splice junction sites with MapSplice that has higher false positive rates than other tools. Based on these results, the reliability of splice junction site results from MapSplice is smaller than the two tools. We confirm that VETic can test reliability to splice junction detecting tools.

4   Conclusions

NGS technology has many advantages over EST and microarray technology. RNA-Seq has the advantage that genes of unknown species may also be applied. Novel splice junctions were difficult to find using EST or microarray technology. However, using RNA-Seq, *ab initio* junction detection has become possible.

In recent years, the Burrows-Wheeler Transform[7] and bioinformatics have been integrated. Alignment tools, such as Bowtie[8], BWA[9], BWA-SW[10], have also increased alignment speed. Several splice junction detection tools have been developed due to these technologies. However, each tools results differ, because they use their own algorithm. Therefore, it is hard to configure which tools have better reliability, since there is not a standard method to verify these yet.

The reliability of the resulting value can be obtained uniformly by VETic. We used TopHat, SpliceMap and MapSplice in the experiments, which show that the reliability is higher for SpliceMap, because the number of candidate splice junction site of latest tool is smaller than for the older tools. In addition, the reliability of tools to have higher false positive rate, like MapSplice, is smaller, because the number of candidate splice junction site is higher than for the other tools.

In the future we plan to a show graphical representation of splice junctions by using this sequence information of the corresponding splice junction site.

References

1. Adams, M., Kerlavage, A., Fields, C., Venter, J.: 3,400 new expressed sequence tags identify diversity of transcripts in human brain. Nature Genetics 4(3), 256–267 (1993)
2. Au, K., Jiang, H., Xing, L.L.Y., Wong, W.: Detection of splice junctions from paired-end rna-seq data by splicemap. Nucleic Acids Res. 38(14), 4570–4578 (2010)
3. Blencowe, B.: Alternative splicing: New insights from global analysis. Cell 126(1), 37–47 (2006)
4. Dimon, M., Sorber, K., DeRisi, J.: Hmmsplicer: A tool for efficient and sensitive discovery of known and novel splice junctions in rna-seq data. PLoSONE 5(11), e13875 (2010)
5. Huang, S., Zhang, J., Li, R., Zhang, W., He, Z., Lam, T., Peng, Z., Yiu, S.: Soapsplice: genome-wide *ab initio* detection of splice junctions from rna-seq data. Frontiers in Genetics 2(0), 46 (2011)
6. Jiang, H., Wong, W.: Seqmap: mapping massive amount of oligonucleotides to the genome. Bioinformatics 24(20), 2395–2396 (2008)
7. Lam, T., Sung, W., Tam, S., Wong, C., Yiu, S.: Compressed indexing and local alignment of dna. Bioinformatics 24(6), 791–797 (2008)
8. Langmead, B., Trapnell, C., Pop, M., Salzberg, S.: Ultrafast and memory-efficient alignment of short dna sequences to the human genome. Genome Biology 10(3), R25 (2009)
9. Li, H., Durbin, R.: Fast and accurate short read alignment with burrows-wheeler transform. Bioinformatics 25(14), 1754–1760 (2009)
10. Li, H., Durbin, R.: Fast and accurate long-read alignment with burrows-wheeler trnasform. Bioinformatics 26(5), 589–595 (2010)
11. Marioni, J., Mason, C., Mane, S., Stephens, M., Gilad, Y.: Rna-seq: An assessment of technical reproducibility and comparison with gene expression arrays. Genome Res. 18(9), 1509–1517 (2008)

12. Pruitt, K., Tatusove, T., Maglott, D.: Ncbi reference sequences (refseq): a curated non-redundant sequence database of genomes, transcripts and proteins. Nucl. Acids Res. 35(suppl. 1), 61–65 (2007)
13. Quinlan, A., Hall, I.: Bedtools: a flexible suite of utilities for comparing genomic features. Bioinformatics 26(6), 841–842 (2010)
14. Trapnell, C., Pachter, L., Salzberg, S.: Tophat: discovering splice junctions with rna-seq. Bioinformatics 25(9), 1105–1111 (2009)
15. Wang, K., Singh, D., Zeng, Z.e.a.: Mapsplice: Accurate mapping of rna-seq reads for splice junction discovery. Nucleic Acids Res. 38(18), 178 (2010)
16. Yen, G., Holste, D., Kreiman, G., Burge, C.: Variation in alternative splicing across human tissues. Genome Biology 74, R74 (2004)

UNION: An Efficient Mapping Tool Using UniMark with Non-overlapping Interval Indexing Strategy

Che-Lun Hung[1], Chun-Yuan Lin[2], and Yu-Chen Hu[3]

[1] Dept. of Computer Science & Communication Engineering, Providence University
200 Chung Chi Rd., Taichung 43301, Republic of China (Taiwan)
clhung@pu.edu.tw
[2] Dept. of Computer Science & Information Engineering, Chang Gung University
259 Wen-Hwa 1st Road, Kwei-Shan Tao-Yuan 333, Republic of China (Taiwan)
cyulin@mail.cgu.edu.tw
[3] Dept. of Computer Science & Information Management, Providence University
200 Chung Chi Rd., Taichung 43301, Republic of China (Taiwan)
ychu@pu.edu.tw

Abstract. NGS has become a popular research field in biologists because it was able to produce inexpensive and accuracy short biology sequences very fast. NGS technique has been improved to produce long length sequences, more than 100bp, recently with the same quality, accuracy and speed. Thus, tools for short sequences may be not suitable for long length sequences. We propose a new tool called UNION for re-sequencing applications by mapping long length sequences to a reference genome. UNION uses the UniMarker with a non-overlapping interval indexing strategy and a tool, CORAL, to do sequence alignments. For the experiments we randomly cut ten thousands sequences with a length of 512bp from the genome of Trichomonas and also produce mutations/sequence errors for these sequences to simulate different similarities. UNION has been compared with GMAP in terms of speed and accuracy and achieves better performance than that of GMAP.

Keywords: NGS, re-seqencing, UniMaker, Genome mapping.

1 Introduction

Recently, with rapid growth of next-generation sequencing technologies (NGS), such as ABI-SOLiD, Roche-454 and Illumina-Solexa systems, have been developed to promptly produce more than 1,000 million short reads, and thus more and more genomic DNAs can be sequenced easily. These genomic DNAs are able to be used to characterize human genetic variations, and the demand for a fast and accurate method to genomically position these DNA sequences is desired. This is traditionally achieved by aligning numerous short DNA sequences, each typically of a few hundred nucleotides, one at a time with one very long DNA sequence. For example, to ensure that a Single Nucleotide Polymorphism (SNP) sequence is mapped to its cognate genomic position, the protocol published by the Whitehead

T.-h. Kim et al. (Eds.): DTA/BSBT 2011, CCIS 258, pp. 187–196, 2011.

Institute uses a double-BLAST [1] search strategy with stringent match criteria (`http://snp.cshl.org/data/`). By focusing on near-identity matches, faster DNA sequence aligning programs have also been developed [2-6]. However BLAST ignores the gap between local alignments, sometimes BLAST is unable to obtain the optimal resulting alignment.

Chen *et al.*, [7] proposed a mapping method that lies in dispensing with the need to perform actual alignment for positional mapping; instead, fixed length unique sequence markers, referred to as UniMarkers or UMs, were used to assign the genomic positions of SNP sites. By definition, every UM appears only once in the genome. Consequently, in the ideal situation of, for example, no sequence errors, a single UM match will suffice to locate an SNP sequence in the genome. The problem of UM method is that memory requirement and computational cost are very huge in case the UMs are long. MUGUP [8] is developed to decrease the computing time. The time of comparing shorter UM is less than comparing longer UM but success rate of mapping is lower. Actually, short UM is included in long UM. Therefore, MUGUP adopts the short UMs and long UMs to build multi-layer marker table. The size of short UMs is less than that of long UMs. In MUGUP, it queries the short UM table first and then find the long UM from the long UM table by the index from short UM table. Hence, it can reduce the computing time using short UM table and keep the same success rate of mapping rate as using long UM. However, this method still suffers from large memory usage and computing time. The tables constructed with different UM lengths increases the memory requirement. If the table is bigger than the memory capacity, it will be stored in the hard driver. Therefore, the comparing time is increasing as querying the table stored in hard driver. In addition, UM coverage rate is the fact of the success rate of mapping. If UMs only locate on few regions of a genome, UMs cannot cover whole region of a genome. Hence, the query sequences only can map to the genome in these regions that have UMs.

GMAP [9] has been proposed for mapping and aligning cDNA sequences to a genome. It includes an overlapping strategy for genomic mapping, oligomer chaining for approximate alignment, sandwich dynamic programming for splice site detection, and microexon identification with statistical significance testing. It adopts overlapping sampling intervals to build the index table. Therefore, it produces many candidate mapping results. For some applications, it is not useful for biologists. In addition, it consumes huge computing time to perform alignment because it adopts dynamic programming approach.

We propose a new mapping and aligning method, UNION, which combines UM [7] and non-overlapping sliding window approaches [10] for genomic mapping and a fast alignment method to solve the problem of low UM coverage rate and decrease the computing time of alignment. Then, a fast and accurate alignment tool, CORAL [11], is used to align markers and short sequences rapidly. Therefore UNION can position short sequences on genome accurately and fast. From the experiment results, UNION achieves better performance than that of GMAP in term of mapping accuracy and computational time.

2   Method and Materials

2.1   UniMarker

UniMarker is an N-mer DNA sequence that is unique in a genome with a sufficiently large value of N. An N-mer sliding window is moved down the genome sequence one base at a time to find all N-mers that occur only once in the genome. These unique N-mers are regarded as UniMarkers. The sliding window methods can be classified into two classes: overlapping and non-overlapping. In overlapping method, the current N-mer sliding window will overlap the previous N-mer sliding window by K-mer where $K < N$. In contrast with overlapping, the current N-mer sliding window does not overlap the previous N-mer sliding window. Figure 1 shows these two methods.

The amount of the UMs is related to the length of UM. The longer length can lead to produce more UMs [9]. However, this relationship is not direct proportion when the length is over a specific value. In their experiments, the amount of UMs produced by 21-mer is seven times over that by 14-mer, and the amount of unique-markers produced by 28-mer is similar to that by 21-mer [7, 8]. In our observations, relation between the length and amount also exists in Arabidopsis Thaliana and Trichomonas Vaginalis.

Low UM coverage rate and UM cluster lead to the query sequence cannot be located at the genome. Due to the low UM coverage rate, some regions has no UM in it that called UM desert. Figure 2 shows the UM coverage and UM desert. The short read cannot map to UM desert by UM method. Besides, the amount of UM is variety of different species. As in Arabidopsis Thaliana, the UM coverage rate is very high. But the UM coverage rate is low in Trichomonas Vaginalis. In the table 3, the coverage rate is only 40% at UM-28 where the length of UM is 28. If position of a query sequence on the genome is located in a region that has no UM there, this query sequence cannot be mapped on the genome sequence accurately.

2.2   UNION

To solve the problem caused by low coverage rate, we propose the method, called UNION, which combines the advantages of UM and non-overlapping sliding window approach. Though non-overlapping sliding window approach, every N-mer can be regarded as a marker in UM desert area. Since BLAT has proved that the non-overlapping sliding window approach is useful to map similar species, we use it to abstract all markers from the genome. UNION adopts two hash tables; UM hash and non-overlapping hash tables. UM hash table is used to record all UMs and non-overlapping hash table is used to record all markers by performing non-overlapping sliding window approach. UNION consists of three stages: (i) building UM hash and non-overlapping hash tables; (ii) mapping querying sequence to UM and non-overlapping hash tables; (iii) aligning query sequence and candidate UMs and finding the final UMs. Figure 3 illustrates the procedure of UNION.

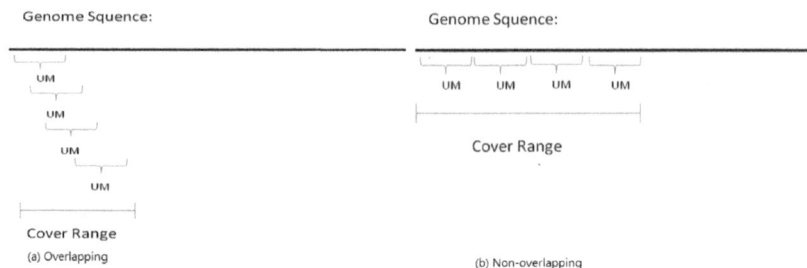

Fig. 1. Sliding window approaches for UM; overlapping and non-overlapping

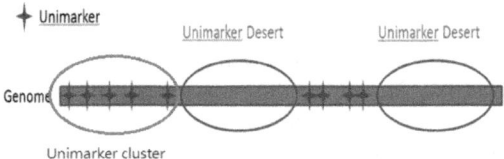

Fig. 2. UM coverage-UM cluster and UM desert

2.2.1 UM Hash and Non-overlapping Hash Tables

From the previous lectures [7, 8], they indicated that the amount of UM-21 is similar to amount of UM-28 and the UM coverage rate of UM-21 is similar to that of UM-28. In addition, the memory usage of UM hash table by UM-28 is bigger than that by UM-21. When the memory usage of hash table is bigger than main memory capacity, the hash table should be stored in hard driver and the efficiency of searching hash table is impaired by searching hard driver. To save the memory usage of UM hash table, UNION adopts 21-mer to build UM hash table.

In UNION, the nucleotides A, T, G and C are encoded as two bits binary data. A, T, G and C are encoded as 00, 01, 10 and 11, respectively. A 16 bits integer can store seven nucleotides and 21 nucleotides can be stored in three 16 bits integers. UNION adopts overlapping sliding window approach to build UM hash table. Every 21 nucleotides form a marker from first nucleotides to last (k-21) nucleotides sequentially; as position 1 to position 21, position 2 to position 22 and so on. After all markers are found, the unique marker can be defined as a UniMarker. These UniMarkers thus are stored in UM hash table. Different to UM hash table, the non-overlapping hash table is build by the markers which are not overlapping and the markers are formed by 1-21, 22-43, and so on. Figure 4 present the markers are abstracted by non-overlapping sliding window from UM desert.

For 32-bit and 64-bit computers, a pointer data structure needs 4 bytes and 8 bytes to store data, respectively. Therefore, the memory usage of UM hash table is 12×4^k bytes for 64-bit computer, where k is the length of UM. For saving the memory, the index table of UniMarkers can be reduced as shown in Fig. 5. In the Fig. 5, the length

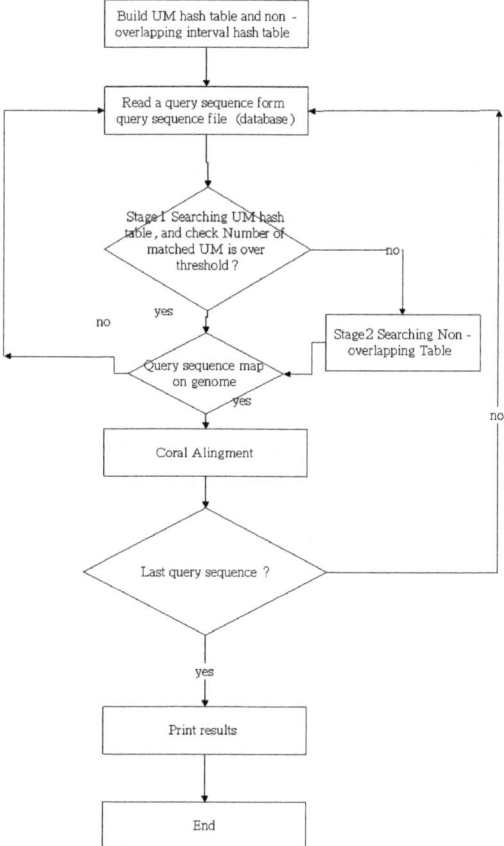

Fig. 3. Procedure of UNION

of UM is 21 and the length of index markers is 14. Therefore, the memory usage of UM hash table can be reduced. However, the number of UMs affects the searching speed on UM hash table. It is trade-off between memory usage and searching speed for UM hash table.

Similar to UM hash table, non-overlapping hash table is a three-layer table. The fist layer index table is same as UM hash table. The second layer table saves the all non-duplicated 21-mer markers found by non-overlapping sliding window approach. The third layer table records the positions of markers on genome sequence. The data structure of non-overlapping hash table is shown in Fig. 6. In our observation, the total amount of memory usage is 7G bytes for building UM hash table and non-overlapping hash table for Trichomonas Vaginalis. Since the first layer table of UM hash table is the same as non-overlapping hash table, these two tables use the same first layer table to save the memory usage. By this way, the total amount of memory usage can be reduced to 4G bytes.

2.2.2   Marker Searching and Sequence Alignment

After two hash tables have been built, the next step is to compare query sequence and markers. The query sequence can be decomposed as $(k-21)$ sub-query sequence by non-overlapping approach with 21-mer, where k is length of query sequence. First is to search UM hash table. When the sub-query sequence is the same as a UM found from UM hash table, it presents that this sub-query sequence can be located correctly at the position of the UM on genome sequence. The mutation and repeat of sequence can lead the mapping error. To avoid the mapping error, the number of matched UMs should be greater than a threshold. But, the mapping accuracy is impaired by the high threshold. Actually, the threshold is related to the length of query sequence. UNION focuses on the long query sequence over 250 bp and the long query sequence can improve the accuracy of mapping of genome sequence with many repeats. The genome sequence of Trichomonas Vaginalis has many long repeats. Therefore it is suitable for our experiments.

If the number of matched UMs is less than the threshold, the query sub-sequences are compared with the markers in non-overlapping hash table. After all matched markers have been found, the diagonal of matched markers is a candidate range if this diagonal is the same as that of matched UMs by the position. By the BLAT perfect match function, the probability of a 21-mer marker existing in a query sequence with 250 bp is 99.99%. UNION adopts BLAT perfect match function to decide the candidate range. The number of perfect match is the factor of alignment accuracy, and this factor will be discussion in experiment section. When all candidate ranges are found, UNION adopts CORAL [11] to compare query sequence and all candidate ranges sequentially. Then, the candidate ranges with higher similarity to query sequence can be found. The query sequence can be located to the position on genome sequence by these matched UMs and candidate ranges.

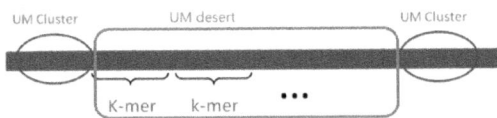

Fig. 4. The k-mer abstraction from UM desert by k-mer non-overlapping sliding window

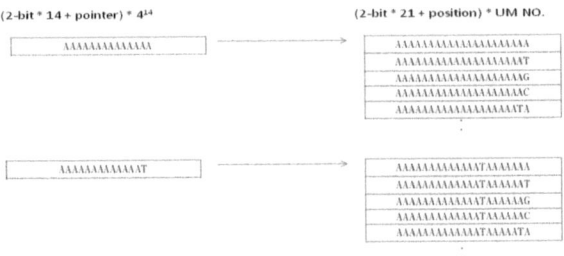

Fig. 5. Data structure of UM hash table

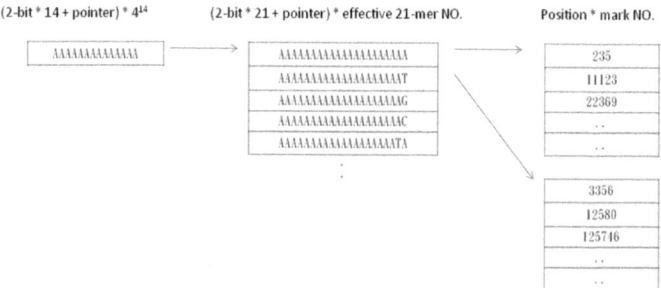

Fig. 6. Data structure of non-overlapping interval hash table

3   Experiment

3.1   Experimental Environment and Data Source

UNION is proposed to compensate the shortcomings of UM approach [7]. The target genome sequence we used in the experiment is Trichomonas Vaginalis sequence that has a lot of repeats. We randomly cut 10,000 sub-sequences with 512 bp as query sequences from exon of Trichomonas Vaginalis sequence.

3.2   Experimental Results

Table 1 presents the relation among sequence similarity, the number of perfect match (PM) and success rate of mapping. In this table, similarity denotes the similarity between query sequences, PM denotes the threshold of perfect match, and total rate denotes the success rate of mapping of UNION. From table 1, UM method only can achieve the accuracy of mapping around 60%. But UNION can achieve the accuracy of mapping that is above 95%. It also indicates that the number of perfect match can affect the accuracy of mapping; especially in lower similarity in the experiment (91% and 90%).

Table 2 presents the relation among sequence similarity, the number of perfect match, success rate of mapping, the number of candidate ranges and the number of candidate ranges with the high similarity scores over 90 by CORAL. It presents that the number of perfect match is contrary to the number of candidate ranges and lower similarity leads to less candidate ranges. However, less candidate ranges can descries the number of performing CORAL to align candidate ranges. UNION only takes 1,033 seconds to process ten thousand query sequences that are 512 bp.

Table 3 presents the comparison of mapping accuracy and computational time of UNION and GMAP. In this experiment, the perfect match is 3. Both of UNION and GMAP use the five candidate ranges with first five highest similarity scores. It is

Table 1. Relation among sequence similarity, the number of perfect match (PM) and success
rate of mapping

Similarity	PM	UM	Stage2	Total rate(%)
99	1	6775	3220	99.95
99	2	6775	3220	99.95
99	3	6775	3220	99.95
98	1	6725	3269	99.94
98	2	6725	3269	99.94
98	3	6725	3269	99.94
97	1	6662	3330	99.92
97	2	6662	3330	99.92
97	3	6662	3330	99.92
96	1	6636	3354	99.9
96	2	6636	3354	99.9
96	3	6636	3354	99.9
95	1	6592	3390	99.82
95	2	6592	3390	99.82
95	3	6592	3390	99.82
94	1	6545	3436	99.81
94	2	6545	3436	99.81
94	3	6545	3433	99.78
93	1	6484	3490	99.74
93	2	6484	3488	99.72
93	3	6484	3472	99.56
92	1	6422	3543	99.65
92	2	6422	3530	99.52
92	3	6422	3457	98.79
91	1	6374	3592	99.66
91	2	6374	3539	99.13
91	3	6374	3343	97.17
90	1	6306	3622	99.28
90	2	6306	3520	98.26
90	3	6306	3197	95.03

obvious that GMAP achieves lower success mapping rates in lower sequence
similarities. UNION can achieve better success mapping rates over GMAP in all
experimental sequence similarities. Also, UNION is faster than GMAP; especially in
high sequence similarity since many repeats need to be aligned with candidate
ranges.

4   Conclusion

The technologies for mapping and aligning cDNA to a genome sequence have been in
development for many years. Each of these tools can be used to satisfy the different
targets. For the short sequences, the alignment is usually unnecessary because they can
be mapped directly without alignment. Therefore, a number of sequences can be
processed rapidly. BLAST is suitable to align and map short sequences to a genome

sequence, especially in comparing different species and finding more similar fragments. For comparing high similar species, BLAT can compare data fast, but the limitation is that it cannot process diverse data, especially different species. GMAP is a useful tool that can solve the limitation of BLAT and is faster than BLAT. However, these tools only produce the results that the similar regions of genome sequence where the short sequences are located possible. The biologists have to decide which similar regions are correct to get the correct positions where short sequence can map. For this problem, UNION is an alternative tool to solve it. UNION can rapidly find the accurate mapping positions on genome sequence where the short sequence can map.

UNION combines UM, BLAT and CORAL to achieve the requirement of re-sequencing that is fast and accuracy. UM method is a useful tool to position the short sequence on genome sequence by UMs. However UM method cannot find accurate positions when UM desire zone. UNION adopts the non-overlapping approach as BLAT to produce markers in UniMark desire zone. Therefore, no marker is missed by using UM and BLAT. Then, CORAL can be used to align markers and short sequences rapidly. Therefore UNION can position short sequences on genome accurately and fast.

Table 2. Comparison of relation among sequence similarity, the number of perfect match (PM), success rate of mapping (total rate), the number of candidate ranges and candidate ranges with high score

Similarity	PM	Total rate	Candidate	HighScoreCandidate	Time(sec)
99	1	99.95	3125830	859370	2502.21
99	2	99.95	3031458	858844	2469.74
99	3	99.95	2920129	856566	2358.09
98	1	99.94	3051293	834891	2445.83
98	2	99.94	2947297	833768	2351.78
98	3	99.94	2816529	829485	2365.22
97	1	99.92	2982769	816573	2350.73
97	2	99.92	2862709	814256	2360.69
97	3	99.92	2700729	805843	2253.4
96	1	99.9	2883476	791103	2343.07
96	2	99.9	2743217	785984	2220.6
96	3	99.9	2542234	769653	2183.22
95	1	99.82	2774524	764486	2212.68
95	2	99.82	2606084	753385	2198.76
95	3	99.82	2354271	722065	2015.98
94	1	99.81	2661006	731127	2192.04
94	2	99.81	2459227	711225	2037.74
94	3	99.78	2146605	659458	1901.82
93	1	99.74	2541361	685318	2063.26
93	2	99.72	2300008	654194	1973.61
93	3	99.56	1911629	577393	1665.38
92	1	99.65	2385364	609901	1995.03
92	2	99.52	2089223	560319	1748.09
92	3	98.79	1624640	455873	1422.3
91	1	99.66	2231722	523873	1823.06
91	2	99.13	1895380	463274	1620.15
91	3	97.17	1378029	348537	1165.17
90	1	99.28	2105169	424069	1734.6
90	2	98.26	1756238	362445	1434.08
90	3	95.03	1215846	253463	1033.36

Table 3. Comparison between UNION and GMAP

similarity	time (GMAP)	success rate(GMAP)	time(Uni)	success rate(Uni)
99	19030	96.38	2358.09	99.95
98	10067	94.72	2365.22	99.94
97	5570	93.35	2253.4	99.92
96	4261	91.73	2183.22	99.9
95	2835	89.76	2015.98	99.82
94	2225	88.44	1901.82	99.78
93	1723	86.12	1665.38	99.56
92	1778	84.44	1422.3	98.79
91	1377	82.66	1165.17	97.17
90	1315	81.08	1033.36	95.03

Acknowledgment. This research was partially supported by the National Science Council under the Grants NSC-100-2221-E-126 -007 -MY3.

References

1. Altschul, S.F., Gish, W., Miller, W., Myers, E.W., Lipman, D.J.: Basic local alignment search tool. J. Mol. Biol. 215, 403–410 (1990)
2. Chao, K.M., Zhang, J., Ostell, J., Miller, W.: A tool for aligning very similar DNA sequences. Comput. Appl. Biosci. 13, 75–80 (1997)
3. Zhang, J., Madden, T.L.: PowerBLAST: A new network BLAST application for interactive or automated sequence analysis and annotation. Genome Res. 7, 649–656 (1997)
4. Florea, L., Hartzell, G., Zhang, Z., Rubin, G.M., Miller, W.: A computer program for aligning a cDNA sequence with a genomic DNA sequence. Genome Res. 8, 967–974 (1998)
5. Delcher, A.L., Kasif, S., Fleischmann, R.D., Peterson, J., White, O., Salzberg, S.L.: Alignment of whole genomes. Nucleic Acids Res. 27, 2369–2376 (1999)
6. Zhang, Z., Schwartz, S., Wagner, L., Miller, W.: A greedy algorithm for aligning DNA sequences. J. Comput. Biol. 7, 203–214 (2000)
7. Chen, L.Y.Y., Lu, S.H., Shih, E.S.C., Hwang, M.J.: Single nucleotide polymorphism mapping using genome-wide unique sequences. Genome Res. 12, 1106–1111 (2002)
8. Hsu, F.R., Chen, J.F.: Aligning ESTs to Genome Using Multi-Layer Unique Makers. In: Proceedings of the Computational Systems Bioinformatics, CSB 2003 (2003)
9. Wu, T.D., Watanabe, C.K.: GMAP: a genomic mapping and alignment program for mRNA and EST sequences. Bioinformatics 21(9), 1859–1875 (2005)
10. James Kent, W.: BLAT—The BLAST-Like Alignment Tool. Genome Research 12, 656–664 (2002)
11. Hung, C.L., Lin, C.Y., Chung, Y.C., Hsieh, S.J., Tang, C.Y.: Comparative Exon Prediction based on Heuristic Coding Region Alignment. In: Proceedings of the International Symposium on Parallel Architectures, Algorithms, and Networks, pp. 14–19 (2005)

Author Index

GPSR Compliance

The European Union's (EU) General Product Safety Regulation (GPSR) is a set of rules that requires consumer products to be safe and our obligations to ensure this.

If you have any concerns about our products, you can contact us on ProductSafety@springernature.com

In case Publisher is established outside the EU, the EU authorized representative is:

Springer Nature Customer Service Center GmbH
Europaplatz 3
69115 Heidelberg, Germany

Batch number: 09474016

Printed by Printforce, the Netherlands